Cosmopolitan Urbanism

'*Cosmopolitan Urbanism* brings together the most incisive writing to date on the idea of the cosmopolitan. Trans-disciplinary throughout, it is a vital contribution to its field.'

Malcolm Miles, University of Plymouth

'Cosmopolitanism, a major pursuit of contemporary interdisciplinary scholarship, rests upon geographical properties that shape both its definition and its achievement. This fine collection of papers successfully demonstrates how space, scale and urbanism are central dimensions of cosmopolitan societies.'

David Ley, University of British Columbia

How are people managing to live with difference in the contemporary city? In grappling with this question, discourses of the 'cosmopolitan city' have become commonplace in debates about contemporary urban development. However, what does cosmopolitanism mean when grounded in the city? And what effect is this having on urban development and how difference is treated or lived with in the contemporary city? This book examines how cosmopolitanism is being interpreted and applied in a range of contemporary urban settings, to critically interpret what cosmopolitan urbanism means in practice and how its deployment is shaping the treatment of difference in the city.

Cosmopolitan Urbanism draws together the work of leading urban scholars to explore the everyday practices of cosmopolitanism and to attempt to answer such questions as: which forms of cultural difference are valued and which are excluded from this re-visioning of the contemporary city?

This groundbreaking book examines the complex politics of cosmopolitanism as it is grounded in the urban, through accessible, empirical case studies that range from Montreal to Singapore, London to Texas, Auckland to Amsterdam. This approach to the discussion of cosmopolitanism makes *Cosmopolitan Urbanism* an accessible student guide to debates on the subject.

Jon Binnie is Reader, **Julian Holloway** is Lecturer and both **Steve Millington** and **Craig Young** are Senior Lecturers in Human Geography at Manchester Metropolitan University.

WITHDRAWN
UTSA LIBRARIES

Cosmopolitan Urbanism

Edited by
Jon Binnie, Julian Holloway,
Steve Millington and Craig Young

LONDON AND NEW YORK

First published 2006
by Routledge
2 Park Square, Milton Park, Abingdon, Oxon OX14 4RN

Simultaneously published in the USA and Canada
by Routledge
270 Madison Ave, New York, NY 10016

Routledge is an imprint of the Taylor & Francis Group

© 2006 Jon Binnie, Julian Holloway, Steve Millington and Craig Young

Typeset in Times by
Book Now Ltd
Printed and bound in Great Britain by
TJ International Ltd, Padstow, Cornwall

All rights reserved. No part of this book may be
reprinted or reproduced or utilised in any form or by any
electronic, mechanical, or other means, now known or hereafter
invented, including photocopying and recording, or in any
information storage or retrieval system, without
permission in writing from the publishers.

British Library Cataloguing in Publication Data
A catalogue record for this book is available from the British Library

Library of Congress Cataloging in Publication Data
Cosmopolitan urbanism/edited by Jon Binnie . . . [et al.].
p. cm.
Includes bibliographical references and index.
1. Sociology, urban. 2. Cosmopolitanism. 3. City dwellers. 4. Ethnic relations. 5. Race relations. 6. Urban geography. 7. Pluralism (Social sciences) 8. Spatial behavior. 9. Cities and towns–Case studies.
I. Binnie, Jon.
HT119.C68 2006
307.76–dc22 2005013886

ISBN10: 0–415–34491–3 (hbk)
ISBN10: 0–415–34492–1 (pbk)

ISBN13: 9–78–0–415–34491–3 (hbk)
ISBN13: 9–78–0–415–34492–0 (pbk)

Contents

List of illustrations vii
List of contributors viii
Acknowledgements xii

1 Introduction: grounding cosmopolitan urbanism: approaches, practices and policies 1
JON BINNIE, JULIAN HOLLOWAY, STEVE MILLINGTON AND CRAIG YOUNG

PART I
Envisaging cosmopolitan urbanism 35

2 Cosmopolitan urbanism: a love song to our mongrel cities 37
LEONIE SANDERCOCK

3 The paradox of cosmopolitan urbanism: rationality, difference and the circuits of cultural capital 53
GARY BRIDGE

4 Strangers in the cosmopolis 70
KURT IVESON

PART II
Consuming the cosmopolitan city: materialities and practices 87

5 **Sociality and the cosmopolitan imagination: national, cosmopolitan and local imaginaries in Auckland, New Zealand** 89
ALAN LATHAM

6 **Cosmopolitanism by default: public sociability in Montréal** 112
ANNICK GERMAIN AND MARTHA RADICE

7 **Cosmopolitan camouflage: (post-)gay space in Spitalfields, East London** 130
GAVIN BROWN

8 **Negotiating cosmopolitanism in Singapore's fictional landscape** 146
SERENE TAN AND BRENDA S. A. YEOH

PART III
Producing the cosmopolitan city: cultural policy and intervention 169

9 **Multicultural urban space and the cosmopolitan 'Other': the contested revitalization of Amsterdam's Bijlmermeer** 171
ANNEMARIE BODAAR

10 **Working-class subjects in the cosmopolitan city** 187
CHRIS HAYLETT

11 **Planning Birmingham as a cosmopolitan city: recovering the depths of its diversity?** 204
WUN FUNG CHAN

12 **Cosmopolitan knowledge and the production and consumption of sexualised space: Manchester's Gay Village** 220
JON BINNIE AND BEVERLEY SKEGGS

13 **Conclusion: the paradoxes of cosmopolitan urbanism** 246
JON BINNIE, JULIAN HOLLOWAY, STEVE MILLINGTON AND CRAIG YOUNG

Index 254

Illustrations

Figures

5.1 Armadillo 90
5.2 Karangahape Road 91
5.3 An ordinary Wednesday, July 1999 99
5.4 Glasses 105
5.5 Jervois Road 106
5.6 Tuatara 107
6.1 Immigration by source area to metropolitan regions, 2001 117
6.2 The seven multiethnic neighbourhoods selected for our study 119

Tables

6.1 Top ten countries of origin of immigrants to the metropolitan region of Montréal during the 1990s 116
9.1 Population of Southeast by ethnicity, 2000 178

Contributors

Jon Binnie is a Reader in Human Geography at the Manchester Metropolitan University. His research interests include the geographies of sexual politics and cultures in an urban and transnational context. He is the author of *The Globalization of Sexuality* (Sage, 2004), *The Sexual Citizen: Queer Politics and Beyond* (co-authored with David Bell; Polity, 2000) and *Pleasure Zones: Bodies, Cities, Spaces* (co-authored with David Bell, Ruth Holliday, Robyn Longhurst and Robin Peace; Syracuse University Press, 2001). He is currently completing a book on the spatialities of citizenship for Routledge.

Annemarie Bodaar is a Ph.D. candidate at Ohio State University and a researcher at the Institute for Migration and Ethnic Studies at the University of Amsterdam. Her research interests include questions of social identity and difference, immigration, and social policy in the context of European cities. She is currently starting a project on ethnically diverse neighbourhoods as spaces of consumption.

Gary Bridge is a Professor at the Centre for Urban Studies, School for Policy Studies, University of Bristol. His recent publications include *Reason in the City of Difference* (Routledge, 2005), *Gentrification in a Global Context* (co-edited with Rowland Atkinson; Routledge, 2005), *A Companion to the City* (co-edited with Sophie Watson; Blackwell, 2000) and *The Blackwell City Reader* (co-edited with Sophie Watson; Blackwell, 2002).

Gavin Brown is in the final stages of a Ph.D. on 'The Production of Gay and Queer Space in East London' in the Department of

Geography at King's College London. He has published several pieces on sexual geography and is currently embarking upon a study of widening participation strategies for medical education in London.

Wun Fung Chan is Lecturer in Human Geography at the University of Strathclyde. He teaches and writes in the areas of urban studies and cultural geography, with a particular interest in the relationship between multiculturalism and city building. His work has appeared in *Society and Space* and the *Journal of Historical Geography.*

Annick Germain is a Professor at the Institut National de la Recherche Scientifique – Urbanisation, Culture et Société in Montréal, Canada. She taught for many years at the Université de Montréal and has written widely on urban planning and immigration. She is co-author with Damaris Rose of *Montréal: The Quest for a Metropolis* (Wiley, 2000).

Chris Haylett is a Lecturer in Geography at the University of Manchester. She is a cultural geographer with research interests in welfare and urban change in Britain and North America. Her publications focus on class cultures and politics, and she is currently working with the ESRC Centre for Research on Socio-Cultural Change in relation to working-class senses of place and value.

Julian Holloway is a Lecturer in Human Geography at the Manchester Metropolitan University. His research interests include practice, embodiment and materiality in the context of urbanism, religion, and countercultural movements. His work has appeared in journals such as *Environment and Planning A* and *Social and Cultural Geography*, as well as books such as *Thinking Space* (co-authored with James Kneale, edited by Mike Crang and Nigel Thrift; Routledge, 2000).

Kurt Iveson is a Lecturer in Urban Geography at the University of Sydney, Australia. He has researched and published on the urban dimensions of citizenship, the geographies of public spheres, and the politics of urban planning.

Alan Latham is a Lecturer at the University of Southampton, UK. His work focuses on mobility, urban sociality, and urban public space. He has published widely in edited collections and international social science journals, including *Progress in Human Geography*, *Environment and Planning A* and *D*, *Area*, *Urban Studies*, and *Urban Geography*.

Steve Millington is a Senior Lecturer in Human Geography at the Manchester Metropolitan University. His research focuses upon local economic development, urban governance, and the cultural economy. His Ph.D., completed in 2002, is entitled 'An Assessment of City Marketing Strategies and Urban Entrepreneurialism in UK Local Authorities'.

Martha Radice is a doctoral student at the Institut National de la Recherche Scientifique – Urbanisation, Culture et Société in Montréal, Canada, where she is researching shopping streets in multiethnic neighbourhoods. Her book *Feeling Comfortable? The Urban Experience of Anglo-Montrealers* was published in 2000 (Presses de l'Université Laval).

Leonie Sandercock is Professor in Urban Planning and Social Policy in the School of Community and Regional Planning at the University of British Columbia and chairs the Doctoral Program. Her current research interests include immigration, citizenship, cultural diversity and integration; participatory planning, democracy, and information and communication technologies; fear and the city, particularly as this relates to 'fear of the Other'; and the importance of stories and storytelling in planning theory and practice. She has just completed her tenth book, *Cosmopolis 2: Mongrel Cities of the 21st Century* (Continuum Books, 2003).

Beverley Skeggs is Professor of Sociology at Goldsmiths College, University of London. She is the author of *Formations of Class and Gender: Becoming Respectable* (Sage, 1997), *Class, Self and Culture* (Routledge, 2004), and *Sexuality and the Politics of Violence* (co-authored with Les Moran; Routledge, 2004).

Serene Tan recently completed a Master's degree in Geography at the National University of Singapore. Her current research examines the relationship between performance art and public space in Singapore.

Brenda S. A. Yeoh is Associate Professor, Department of Geography, National University of Singapore, as well as Research Leader of the Asian Migration Research Cluster and Principal Investigator of the Asian MetaCentre at the University's Asia Research Institute. Her research foci include the politics of space in colonial and postcolonial cities; and gender, migration, and transnational communities.

Craig Young is a Senior Lecturer in Human Geography at the Manchester Metropolitan University. His research interests include issues of changing place identity and urban governance in the context of regional and global restructuring. This includes a focus upon UK cities and the post-socialist world. He has published widely in journals including *Political Geography* and *Area*. He is currently completing a book entitled *Global Geographies of Post-Socialist Transitions* (co-authored with Tassilo Herrschel; Routledge).

Acknowledgements

The origins of this collection lie in a conference session entitled 'The Production and Consumption of Cosmopolitan Geographies', held at the Association of American Geographers' 99th Annual Meeting, New Orleans, March 2003. The editors would like to thank the contributors and the audience of these extremely productive and engaging sessions, and the contributors to this volume for their efforts in meeting deadlines and responding to editorial requests. The editors would like to extend their sincerest thanks to the following people who have, in various ways, made this collection possible: Rod Allman, David Bell, Paul Chatterton, Tim Edensor, John Fitzgerald, Shaun French, Joanna Hodge, Ruth Holliday, Mark Jayne, James Kneale, Zoe Kruze, Sam Millington, Josh and Arthur Millington, Kath Moonan, Andrew Mould, Anna Righton, Louise Saunders, Graham Smith, Sandra Walklate, the BeerCall, Honey, Hale and Tiger.

We would also like to acknowledge *The Sociological Review* for permission to reprint J. Binnie and B. Skeggs (2004) 'Cosmopolitan knowledge and the production and consumption of sexualized space: Manchester's gay village', *Sociological Review*, 52: 39–61.

Introduction: grounding cosmopolitan urbanism

Approaches, practices and policies

Jon Binnie, Julian Holloway, Steve Millington and Craig Young

Introduction

On Friday 21 January 2005, the British newspaper *The Guardian* ran a series of articles exploring and commenting upon London as the 'world in one city' where 'every race, colour, nation and religion' can be found and experienced. The lead journalist behind these articles, Leo Benedictus (2005: 2), had apparently 'spent months travelling across the capital, locating and visiting the immigrant communities that give the city its vibrancy', commenting upon his results that 'London in 2005 is uncharted territory. Never have so many different kinds of people tried living together in the same place before . . . New York and Toronto would contest the cosmopolitan crown, but London's case is strong.'

To supplement the articles the newspaper provided its readers with maps of London depicting particular clusters of ethnicity and religion in the capital. Ethnic, religious and cultural diversity, it seems, are at the heart of what makes a twenty-first-century city 'vibrant'. Moreover, it is this difference that apparently makes a city *cosmopolitan*. The tone of *The Guardian* was certainly celebratory in its vaunting of London's supposed status as the capital of cosmopolitanism. Just over a fortnight later, another British newspaper, *The Independent*, published a very different article concerned with ethnic difference in London. The article, by Jason Bennetto the paper's crime correspondent and entitled 'London's cosmopolitan criminals targeted', reported an interview with Sir John Stevens, the now former Commissioner of the

Metropolitan Police, about his concerns with the growth of criminal gangs within immigrant communities in the capital. Like *The Guardian* article, *The Independent* produced a map of neighbourhoods associated with enclaves of immigrant communities, but instead of praising the opportunities to encounter difference through the consumption of food and different communities, the article highlighted which nationalities and ethnic groups were prevalent in each neighbourhood of the city in terms of the forms of criminality most associated with them – for example, Albanians involved in vice in Soho, Jamaican 'Yardies' involved in the drugs trade in Brixton and Hackney. In the interview, Sir John Stevens argues:

> We talk about these 16 new communities; Kosovans, Kurds, Turks and others. I think it is an area we really need to keep an eye on. [Criminals within] these communities need to understand our system of law. Some come from countries where there is not much respect for the police, for obvious good reason. They need to understand that we police this country by consent. We are a tolerant country, but we will not allow any form of criminality, especially organised crime to take place. I think these people sometimes come from countries that think we are a soft touch – we prove otherwise.

The two articles give sharply differing accounts of attitudes towards encounters with ethnic and national difference in the cosmopolitan city. *The Guardian* article celebrates the possibilities and opportunities of living with difference, while the discussion of the violence associated with criminal gangs located within immigrant communities brings into sharp focus the fears associated with encounters with difference in the city.

In a further analysis of difference in British cities, drawing upon the methodology developed by Richard Florida (2002) in *The Rise of the Creative Class*, the 'think-tank' Demos calculated 'The Boho Britain Creativity Index' (Demos 2003). This index aimed to measure the 'creative potential' of 40 of the UK's largest cities, using three key indicators: the number of patent applications per head; the number of residents who are not categorised as 'white British'; and the number of services provided to the gay and lesbian community in the city. Ethnic and sexualised difference is thus held to be central to the sustainability, creativity and entrepreneurialism of cities. As Demos (2003) states, 'Diversity is vital because it is through combining and colliding new and old that innovation and adaptation occur. The more open a city is to new people, new ideas and new ways of living, the greater its creative metabolism will be.' Based on this justification and criterion it is Manchester, followed by Leicester and London, that heads the list for the 'new bohemians' in the UK (Demos 2003).

What is apparent from these British examples is how notions of the cosmopolitan or diverse city are deployed and mobilised in a variety of different contexts. Moreover, these examples reveal that the 'cosmopolitan city' as phrase and idea has not only entered, but arguably has become sedimented in, the discourses of the public sphere. The cosmopolitan city is therefore being written about and discussed far beyond the walls of academia. Indeed, the idea that certain cities, or at least areas of cities, are somehow more or less cosmopolitan is something increasingly encountered in the everyday vernacular of contemporary Western societies. This common usage in itself is something that is worthy of attention. Yet the critical issues surrounding the generation and realisation of the cosmopolitan city warrant further scrutiny. For example, why is diversity seen as key to the success of the contemporary city? How and why are policymakers and key players in the urban realm drawing upon the cultural diversity of cities to gain the 'crown' of 'the most cosmopolitan city'?

Furthermore, how is the diversity at the heart of the cosmopolitan city encountered and practised? What are the consequences for those groups woven into this 'vibrant' cosmopolitan fabric? What are the consequences for those groups (or cities) deemed non-cosmopolitan? It is our contention in drawing the contributors and case studies together in this collection that only by exploring and grounding questions such as these in contemporary urban contexts can we begin to understand what it is to live in and create the cosmopolitan city. This introduction to *Cosmopolitan Urbanism* provides the reader with an overview of contemporary debates on cosmopolitanism and cosmopolitan urbanism through a discussion of competing definitions of these concepts. We begin by setting out the main characteristics of debates on cosmopolitan identities and subjectivities within social, cultural and political theory. We follow this with a discussion that addresses some of the critical issues raised by articles and surveys like those above by tracing the relationship between cosmopolitanism and the city. Here we focus on key debates concerning gentrification, urban formations of the 'new middle class' and new work and consumption spaces. Finally we outline the structure of the book, and discuss how the individual chapters advance debates on cosmopolitan urbanism.

Approaching cosmopolitanism

Since Immanuel Kant's early conceptualisation of cosmopolitanism (see Reiss 1991; Fine 2003a), the term has been variously deployed and variously contested in different literatures. As Szerszynski and Urry (2002: 469) argue, 'there is no one form of cosmopolitanism; it rather functions as an "empty signifier" . . . having to be filled with specific, and often rather different content, in different situated cultural worlds'. The elusiveness of a simple and coherent definition of the term means that we must tread carefully when deploying it. This is further compounded in that cosmopolitanism is often intimately related to and used interchangeably with other terms that act as diagnoses 'of the age in which we live' (Fine 2003b: 452). For example, cosmopolitanism is used frequently in tandem with transnationalism. Yet for some writers, such as Hiebert (2002), a conceptual distinction between transnationalism and cosmopolitanism is necessary, in that transnational communities may not exhibit or embody the qualities specific to the term cosmopolitanism. As such we must acknowledge its multiple usages, but also be careful in discussing it interchangeably with other (albeit related) terms.

This section thus suggests that there are two main ways in which cosmopolitanism has been conceived and understood. As Hiebert (2002: 211–12) suggests:

> In everyday language, the term is generally applied to places with a marked cultural diversity (for example, 'x is a more cosmopolitan city than y'). Among researchers, cosmopolitanism is often equated with political internationalization, whether in the form of institutions such as the United Nations, or global protest movements.

Therefore cosmopolitanism can firstly be conceptualised through its usage in everyday life, especially as a focus on cultural diversity and difference. In particular, cosmopolitanism is often deployed in terms of a specific attitude towards difference and thus the possession of a set of skills that allow individuals to negotiate and understand cultural diversity. As such, this understanding leads to disagreement over the subjectivity or identity of the cosmopolite. Who is s/he? What are the nature and consequences of these skills, both for those that embody them and those who are marked as different? Can we simply locate the subjectivity of the cosmopolite within the body of an elite, educated Western traveller, or is the cosmopolitan more correctly associated with transnational migrants, refugees or asylum seekers?

The second understanding of cosmopolitanism, which Hiebert associates with 'researchers', concerns a political geography and philosophy of global citizenship. In this understanding, a normative project which favours 'universalistic standards of moral judgement, international law and political action' is underpinned by a rejection of citizenship and loyalties based upon the nation and the nation-state (Fine 2003b: 452). This idea of a reframing of notions of citizenship as part of a 'cosmopolitan project' is addressed below.

Cosmopolitan citizenship

Cosmopolitanism is often intimately related to the various processes that exemplify globalisation. Thus as a current condition of society, culture and politics, cosmopolitanism is seen to result from and contribute to the intensification of time-space compression or distanciation. In other words, as we live our lives in increasingly global contexts we have become more aware of the world as both a single place and one comprised of multiple differences. Furthermore, a variety of processes and risks, such as environmental and human rights issues, HIV and AIDS, have transformed the spatial scale of political agency and citizenship. It follows, therefore, that according to advocates of cosmopolitanism, issues of justice, citizenship and politics should now be conceptualised at the global scale. This has consequences for the nation and the nation-state. As Fine (2003b: 453) argues, 'cosmopolitan political philosophy affirms the possibility and desirability of overriding national sovereignty in the name of cosmopolitan justice'. This universalistic vision of justice and political community is characterised by its global frame of reference, and by a reconfiguration of democracy towards a cosmopolitan vision that replaces political organisation and loyalties based on the 'narrow particularisms' of the nation. 'Thinking and feeling beyond the nation', as Cheah and Robbins (1998) have it, is something to be advocated and sought out.

In particular, this vision of cosmopolitanism as a philosophy of world citizenship has been articulated through the notion of cosmopolitan law. Thus, whereas international law recognises the integrity of nation-states and their role in enforcing human rights, cosmopolitan law emphasises individuals and groups as legal personalities. Here cosmopolitanism 'requires the implementation of legal mechanisms that act in the interests of global citizens rather than states' and as such 'under cosmopolitan law the bearers of human rights are individuals and not states or nations' (Stevenson 2002: 2.3, 2.6). Indeed, global socio-economic relations that are arguably undermining the impact of national policies and forms of governance have made developing forms of knowledge, such as cosmopolitan law, increasingly relevant and apposite to the current epoch. For example, Ulrich Beck (2004: 153) goes so far as to say: '[t]he social premises of the national state – a uniform space, nation and state – are no longer present, even though new organizational forms of cosmopolitics are not yet clearly discernible'. However, this view is qualified by Fine (2003b: 452) when he states that the project of cosmopolitics 'aims to reconstruct political life on the basis of an enlightened vision of peaceful relations between nation states, human rights shared by all world citizens, and a global legal order buttressed by a global civic society'.

This perception of a global citizenry acting across and subsequently undermining the sovereignty of the nation and the relevance of national borders is a debatable one. However, it might be that nascent forms of cosmopolitical solidarity are indeed developing. In their essay on everyday practices of global citizenship in the north-west of England, Szerszynski and Urry (2002) identify the development of what they term a 'cosmopolitan civil society' (2002: 477) based on a sense of solidarity with, and responsibility for, 'Others' beyond the boundaries of their own nation. Interestingly, this sense of cosmopolitan civic duty was differently experienced and conceived across the life course, with significant differences in how global rights and responsibilities and senses of being 'at home in the world' were expressed. Thus young people tended to articulate cosmopolitanism through practices of tourism and travel, whereas adults expressed it in terms of responsibilities towards those across the globe, with retired people articulating a new sense of freedom and openness to travel and to experiencing other places and cultures.

A number of criticisms can be levelled against this reading of global civic society in which cosmopolitanism replaces the political, cultural and economic importance of the nation and the nation-state. First, critics argue that the nation-state and sentiments towards the nation are still significant forces in the global economy, polity and society, such that the changes heralded by the emergence of cosmopolitics become 'short-term, transitory or downright illusory' (Fine 2003b: 463). For instance, it is evident from Brenda Yeoh's (2004) study of Singapore that cosmopolitanism is being articulated as a project of nation-building or recreating a national identity for the new millennium based on an economy that is receptive to professional foreign migrant workers who can boost the Singapore economy (see also Tan and Yeoh, this volume). Second, it has been suggested that this articulation of cosmopolitanism may in fact dovetail neatly with the neo-conservative political project in the United States (see, for instance, Kiely 2004). Indeed, it might be that cosmopolitanism is merely another aspect of a wider political, economic and social project of certain globally dominant nation-states whose 'particularistic cultural assumptions, national prejudices and power positions remain intact behind its universalistic discourse and institutions' (Fine 2003b: 464). Third, those groups who could be seen as eschewing loyalties and attachments centred on the nation-state might actually belong to particular transnational communities, where initial impressions of cosmopolitanism are undermined with further investigation. It is to these groups and the dispositions that they are supposed to articulate that we now turn.

Cosmopolitan attitude and practice

Another of the common understandings of the term cosmopolitanism is founded on an openness to, desire for, and appreciation of, social and cultural difference. This definition arises primarily from the work of Ulf Hannerz (1996: 103) and his frequently cited characterisation of cosmopolitanism as 'an orientation, a willingness to engage with the Other . . . [entailing] an intellectual and aesthetic stance toward divergent cultural experiences, a search for contrasts rather than uniformity'. The cosmopolite is therefore open to and actively seeks out the different, in a restless search for new cultural experiences. Cosmopolites reject the confines of bounded communities and their own cultural backgrounds. Instead they are seen to embrace a global outlook. As David Ley (2004: 159) notes, 'Cosmopolitans think globally, aim to exceed their own local specificities, welcome unfamiliar cultural encounters and express the wish to move toward a true humanity of equality and respect, free of racial, national and other prejudices.'

Once again national, and indeed local, particularities are disposed of and new forms of identification based on globality and diversity are sought. Cosmopolitanism is thus an attitude, but also a set of dispositions and forms of practice with and towards diversity. Drawing upon Szerszynski and Urry (2002), we can further explore and characterise this attitude and practice. Cosmopolitan practice involves a set of skills which are applied in the encounter with difference. In particular these skills involve the ability to map one's own socio-cultural position vis-à-vis the diversity encountered, and thus require a degree of reflexive ability whereby the cosmopolite can map and locate such societies and cultures historically and anthropologically. Furthermore, this skilled curiosity towards other societies and cultures often involves a degree of risk by virtue of experiencing diversity and otherness. Thus the skill of the cosmopolite is bound up with moments of uncertainty. Yet this practice remains one where risks are overcome by the ability and willingness of the cosmopolite to make sense of and move through different societies, gathering not only knowledge of the particular culture in question but also enhancing a disposition and attitude that reduces the shock of the new or the different in other circumstances. The cosmopolite therefore becomes skilled in navigating and negotiating difference.

While these characteristics may initially appear uncontroversial, they have been subjected to harsh criticism. In particular, it has been argued that the figure of the cosmopolite must be explored within the context of other identity formations and relations of power. For example, other authors have discussed the gendered nature of cosmopolitanism. Mica Nava (2002) argues that, for British women in the early part of the twentieth century, the desire for the exotic reflected in department stores such as Selfridges may have betrayed a desire to escape and transgress gender and national norms. As she suggests (2002: 85–6), for these women 'cultural difference and the foreign constituted a source of interest, pleasure and counter-identification that existed in tension with more conservative outlooks' and thus they 'appropriate[d] the narratives of difference for themselves in contrary and even polemical ways'.

Another crucial set of identities and power relations which must be taken into account when exploring the figure of the cosmopolite are those of class. In other words, an awareness of transformations in class formations is useful in developing our understandings of cosmopolitanism more generally. Indeed, as Werbner (1999: 18) notes, 'the class dimensions of a theory of global subjectivity have remained largely unexamined'. In particular, cosmopolitanism – as a set of skills or competencies – is an intrinsically classed phenomenon, as it is bound up with notions of knowledge, cultural capital and education: being worldly, being able to navigate between and within different cultures, requires confidence, skill and money. It is no surprise, therefore, that a cosmopolitan disposition is most often associated with transnational elites that have risen to power and visibility in the neoliberal era (Isin and Wood 1999; Sklair 2001; Kennedy 2004). As Ley (2004: 157) describes: 'it is asserted that the erosion of transaction costs and the increasing flexibility of citizenship arrangements have created an "ungrounded" or "deterritorialized" transnational class moving at will' and thus an elite whose members are disembedded and 'globally aligned' (Rofe 2003: 2519).

Despite the fact that these transnational elites exist in positions of power and privilege enabling them to act as cosmopolites, we should be careful not to homogenise the experience of such groups. Put differently, the global alignment of the transnational professional class is more complicated than it might at first seem. For example, as Ley (2004) notes, often these groups are vulnerable to a lack of economic knowledge of the spaces in which they do business, and furthermore global business travel comes with many personal and familial costs. Moreover – and to deepen our understanding of how cosmopolitanism is classed – it is vital to recognise that transnationalism is not an exclusively elite or professional phenomenon. Thus concentrating upon these groups tends to discount other classes and groups for whom transnational and cosmopolitan experiences are a daily reality. In this regard, Werbner (1999: 17) is critical of Hannerz's classed-based framing of the cosmopolitan as a disposition of a Western elite of educated, transnational professionals, arguing that there is an implicit 'separation of professional occupational transnational cultures from migrant or refugee transnational cultures'. Therefore, we must also take into account the existence of a working or low-income transnational class that 'while implying the same processual *forms* of hybridisation and creolisation, do not generate the same cultural hybrids as those evolved by elite cosmopolitans' (Werbner 1999: 23). As such, we must appreciate that cosmopolitanism is classed in multiple ways, and be cautious of equating cosmopolitan values, attitudes and practices exclusively with any one class formation. Differently classed subjects encounter difference and access urban space and move through the city with different compositions of social, cultural and economic capital. For example, there is evidence of cosmopolitanism being produced 'from below' rather than just by transnational elites (and see Smith and Guarnizo 1998). Discussing the democratisation of cosmopolitanism, and challenging definitions that merely associate it with the property of an elite group, Antoine Pécoud (2004: 15) argues that this understanding necessitates an awareness of the complexity of the concept and attention to the specific contexts of its use:

> While acknowledging 'from below' cosmopolitanism is only a matter of doing justice to the complexity of immigrants' lives, it also raises the question of the different forms it may take in different contexts and among different people. Cosmopolitanism may transcend boundaries of class, ethnicity or gender, but is unlikely to remain exactly the same in doing so.

For example, discussing the everyday cosmopolitan practices and skills of German–Turkish entrepreneurs in Berlin, Pécoud (2004: 16) draws attention to the skills and competencies necessary in order to survive and thrive in business: for example, an awareness of the necessity of engagement with different cultures:

> German-Turkish entrepreneurs' cosmopolitanism has both a mental and a practical dimension. Business is a concrete activity that requires actual skills, and in this respect, cosmopolitan competencies are essential to business success. This implies that cosmopolitanism is not always a matter of entrepreneurs' will or pleasure.

So cosmopolitanism is not simply a matter of the individual choice of free agents, but rather, as Craig Calhoun (2002) argues, it is *socially* produced. Pécoud contrasts the disposition of consumers, which functions primarily in terms of taste and a desire for the Other, with that of producers and entrepreneurs whose cosmopolitanism is an economic necessity which is dependent upon obtaining skills which may not always be straightforward, as they may lack education and the capital needed in order to acquire them. There may therefore be an inequality in terms of the resources and labour required to produce a cosmopolitan disposition. At the same time he notes that cosmopolitanism in business is a two-way process, with global capital seeking out ethnic markets as well as ethnic minority entrepreneurs displaying cosmopolitan tendencies and skills in order to widen the markets for their goods and services.

The mobility associated with transnational elites is often derived from, and counterposed against, the immobility of certain groups. In other words, some argue that elite cosmopolites gain their status and ability to develop global attitudes and skills through a pathologisation of those groups who are somehow fixed in place and for whom local or national loyalties still pertain. For example, despite his attempt to think through an embracing view of cosmopolitan virtue, Turner (2000: 141–2) sustains this dualism of mobile versus immobile, articulated in specifically class terms, in the following description:

> Those sections of the population which are relatively immobile and located in traditional employment patterns (the working class, ethnic minorities and the under classes) may in fact continue to have hot loyalties and thick patterns of solidarity. In a world of mounting unemployment and ethnic tensions, the working class and the inhabitants of areas of rural depopulation may well be recruited to nationalist and reactionary parties. Their worldview, rather than being ironic, becomes associated with reactionary nationalism.

In this view there is a devaluing of the locally or nationally bounded sentiments of the working class which is contrasted with the dislocated and highly mobile skills and concomitant attitudes of other more privileged class formations. Indeed, it seems as if it is not only transnational professionals that are given cosmopolitan status. By implication, the middle classes are often held up as cosmopolitan in their attitude and engagement with differences, precisely through being more mobile and less likely to succumb to the immobile attitudes of the 'non-cosmopolitan' classes.

However, this dualism of the mobile cosmopolite versus the immobile non-cosmopolite requires further exploration. It appears that the mobility of the middle classes or elites implies that a dislocation from specific locales or the nation is a condition for achieving cosmopolitan status. In contrast, Kwame Appiah's essay on cosmopolitan patriotism (1998: 91–2) implies that belonging to a particular community might be the necessary underpinning to realising the cosmopolitan society:

> the cosmopolitan patriot can entertain the possibility of a world in which *everyone* is a rooted cosmopolitan, attached to a home of his or her own, with its own cultural particularities, but taking pleasure from the presence of other, different, places that are home to other, different people.

With this in mind we can begin moving beyond the overly simplistic dualism of mobile/immobile to a realisation that holding sentiment for somewhere, rather than eschewing ties to anywhere, might inform and foster a cosmopolitan attitude and disposition to diversity and difference. It follows that 'cosmopolitanism . . . does not necessarily imply an absence of belonging but the possibility of belonging to more than one ethnic and cultural localism simultaneously' (Werbner 1999: 34).

Thus, instead of limiting our focus to the globally oriented worlds of elites, we need to explore and investigate how cosmopolitanism is formed and reformed in particular locales and everyday spaces. We need to examine what Lamont and Aksartova (2002: 1) term 'ordinary cosmopolitanisms', by which they mean 'the strategies used by ordinary people to bridge boundaries with people who are different from them'. As such, Lamont and Aksartova (2002: 2) argue against the notion that cosmopolitanism can be merely associated with elite groups of world travellers:

> Their [ordinary cosmopolites'] absence from this literature implies that cosmopolitanism is less likely to be encountered in their midst as compared to the elite strata of society where 'cosmopolitan travellers' are generally implicitly thought of as dwelling . . . However, this should not be taken to mean that they are bereft of cosmopolitan imagination – they engage with difference perhaps just as often as the paradigmatic cosmopolitans, albeit on a local, as opposed to a global, scale.

Thus the rhetorics deployed and practices performed in local, ordinary and mundane times and spaces are arguably where cosmopolitanism takes its forms and shapes in contemporary societies. For example, a mundane cosmopolitanism for Szerszynski and Urry (2002) is produced through exposure to an ever greater range of exotic and different cultures through TV, and thus primarily in the quotidian space of the home. In this case, therefore, imaginative travel is engendered through media and other forms of communication technologies. Another example of how cosmopolitanism is produced differently through everyday spaces comes from Louisa Schein's (1999) paper on cosmopolitanism and consumption practices in post-socialist China. She argues that the rise of consumer capitalism means that 'media consumers, simultaneously imbibing print, electronic, and satellite communications around the globe, come to imagine themselves as cosmopolitan participants in global commodity culture' (Schein 1999: 345). Schein's paper examines the relationship between class, mobility and desire, coining the term 'imagined cosmopolitanism' to denote attempts to transgress the constraints of locality and nation through interaction with global promotional media, rather than simply through the acquisition of foreign products.

While this desire for the foreign in the form of commodities and material goods is 'about surmounting the spatial constraint of locality, about entering the global scene by means that deny geographic immobilities' (Schein 1999: 345), we can once more identify the spaces of the shopping street and the home as sites for the construction of an imaginative cosmopolitanism and a practised globality. Thus, we can see here the importance of how cosmopolitanism is bound into and expressed through certain commodities and forms of commodification. Indeed, it might be that a focus on ordinary or banal cosmopolitanism reveals how encounters with difference and diversity are themselves becoming articulated through commodities and processes of commodification. As Beck (2004: 150–1) argues:

> Cosmopolitanism has itself become a commodity; the glitter of cultural difference fetches a good price. Images of an in-between world, of the black body, exotic beauty, exotic music, exotic food and so on, are globally cannibalised, re-staged and consumed as produces for mass markets.

Yet such processes need to be examined through particular places and at particular times in order to get a sense of how cosmopolitanism is worked out. In other words, it is through spaces of (in particular) the city that we need to generate an understanding of how these key issues of class, commodification and the everyday intersect with, produce and reveal the attitudes and practices of cosmopolitanism. How 'individuals and groups go about building a world in the city' (Latham 2003: 1703), and how this intersects with class formations, material culture and the quotidian spaces of contemporary urbanism, is something we will turn to next.

Cosmopolitanism and the city

Cosmopolitanism can thus be characterised in two major ways: first, as a philosophy of world citizenship which simultaneously transcends the boundaries of the nation-state and descends to the scale of individual rights and responsibilities in an apparently increasingly connected and globalised world; and, second, as a particular set of skills and attitudes towards diversity and difference. In addition we have highlighted some critical issues that need to be further explored in the articulation of cosmopolitanism. These include the ways in which so-called cosmopolitan identities intersect with and reproduce other socio-cultural distinctions and cleavages, especially class, and how the encounter with difference is bound into processes of commodification and materiality. Furthermore, we have argued that understandings of cosmopolitanism must be generated through analyses which are grounded in particular times and different spatial contexts. In this section, we explore these crucial issues and thus develop a framework for the chapters that follow.

In particular we trace the trajectories of these issues and debates through a focus on the cosmopolis. This section therefore outlines the major debates in urban studies in which the figure of the cosmopolite and the spaces of cosmopolitanism in the city have come to the fore. There are two significant aspects of these debates that we concentrate upon here: first, the ways in which the figure of the cosmopolite and cosmopolitanism have been examined through work on gentrification and the consumption practices of the 'new urban middle class'; and, second, how processes of cosmopolitanism are produced and sustained through different forms of urban governance, and in particular the production of distinct cosmopolitan quarters and sectors through policy and planning. By examining these two debates on the contemporary city we reveal how the encounter with difference and diversity in urban spaces must always be analysed within the context of class, commerce, commodities and political-economic processes which give the cosmopolis a distinct spatiality and temporality. In other words, we examine forms of cosmopolitan urbanism by exploring the kinds of spaces and places within cities where the cosmopolite is located.

Gentrification, distinction and the spaces of the new middle class

To begin answering these questions we must look to debates about gentrification and the so-called 'new' middle classes as one of the main ways in which cosmopolitan urbanism is being investigated and discussed. Thus, as David Ley (2004: 160) notes, it is the 'residents of gentrified inner-city neighbourhoods that have multiple points of openness to cosmopolitanism'. The new middle and gentrifying classes are definable through their particular combination of economic and cultural capital that enables them to distinguish themselves from other classes on a number of fronts. In particular, it is the production and deployment of cultural capital and aesthetic sensibilities that provides the mechanism through which this class fraction finds its identity: the reflexive and conscious production of distinction through forms of cultural capital. Often this cultural capital is materialised through property development and display, and thus 'one of the features of gentrification has been that the deployment of a cultural aesthetic to provide social distinction has in turn enhanced material capital'

(Bridge 2001: 93). Another element in the production of cultural capital, and therefore the distinction of the 'newness' of these middle-class groups, and one which is pertinent to our focus here, is the articulation of a global sense of identity and a practised worldliness of cosmopolitanism.

To exemplify this sense of global identity in the gentrified spaces of the new middle classes, we can look to Rofe's (2003) work on the districts of Glebe in Sydney and Inner Newcastle, Australia. In his study of the gentrification of these areas, 74 per cent of those asked strongly agreed that they were global community members and as such that '"being global" is a title of distinction, infusing the individual with a sense of cosmopolitanism' (Rofe 2003: 2521). Glebe and Inner Newcastle have become residential zones for highly educated professionally employed groups, who consider themselves globally oriented, particularly through artistic and intellectual pursuits which draw them into international cultural circuits. Yet this globally oriented identity is also manifested through the particularity of the urban spaces and contexts in which they are situated: 'being global' 'finds physical expression through local territory . . . sampling revealed a widespread belief that gentrifying landscapes constituted globally oriented places' (Rofe 2003: 2520). In other words, the gentrifying neighbourhood is, for the respondents in Rofe's study, the global grounded in the local. These gentrified urban settings stage a global sense of place, or an aesthetic and cultural globalism. Of particular importance to this grounded urban globality are encounters with difference commodified through ethnic restaurants, import stores, international media and architectural forms. Thus the socialisation of cosmopolitan global identities and the generation of encounters with difference and diversity find their locus in particular 'consumptionscapes' in the city.

It is also important to note how the new middle classes' sense of global and cosmopolitan identity may be formed through a pathologisation of the local. Put differently, 'being global' in the cosmopolis is achieved through a distinction and othering of those for whom the global is not a cultural, social or economic reality or source of identification. Thus Rofe's (2003) respondents demonstrated a practice of detachment from certain forms of localised popular culture. This performance of distinction took a particularly spatial form. The inner city, where gentrified and cosmopolitan identity formations can be made and displayed, is constructed through an othering of the suburb as a space of 'myopic mainstream Australian culture' (Rofe 2003: 2520). Here we see Rofe's respondents making the distinction between a mobile and globally orientated cosmopolite and the immobile and locally fixed non-cosmopolite, who in this particular context is framed as working-class or a 'traditional' suburban formation of the middle class. Therefore, the new middle classes may become exiles in their home culture, through rejecting the national(istic) sentiments of other class positions. Moreover, this process of differentiation may lead to a reduced encounter with certain forms of difference. Indeed, the new middle classes may seek out spaces in the city which are more notable for their homogeneity than their heterogeneity. As Tim Butler (2003: 2469) argues, in the context of gentrification and the new middle class in Barnsbury in Islington, North London, 'in a city which is massively multiethnic, its middle classes, despite long rhetorical flushes in favour of multiculturalism and diversity, huddle together into essentially White settlements in the inner city'. This reflects May's (1996: 196) argument that gentrifiers and the new middle or 'cosmopolitan' classes may in fact be seeking to eradicate rather than celebrate difference:

> Whilst welcoming a world of difference this interest in difference and otherness can also be understood as describing a project of cultural capital through which members of the class seek to display their liberal credentials and thus secure their class position . . . Such a process works to both silence and contain that difference which is sought.

The global habitus of gentrifiers, superficially at least, seems to reflect the attitudes and practices of cosmopolitanism, including an active celebration of and desire for diversity. However, it may in fact produce an exclusion of difference by drawing symbolic boundaries between acceptable and non-acceptable difference. As Ley (2004: 161) suggests, 'the expansiveness of inner-city cosmopolitanism becomes both more partisan and more parochial upon closer examination'. Indeed, this often translates into a defence of the locality through a resistance to land-use change and preservation of certain aspects of local heritage (especially in terms of architecture). There is, then, a paradox within the consumption practices of the new middle-class gentrifier. Far from sharing an open outlook beyond the local, we see a defence of the locality and the jealous guarding of the neighbourhood from encroachment from the outside. Certain groups are not seen as 'appropriate' for the neighbourhood, altering 'the very nature of the cosmopolites from cultural connoisseurs to cultural imperialists' (Rofe 2003: 2525). Ironically this can lead to the loss of a space's distinctive and diverse identity which attracted the globally oriented and cosmopolitan groups in the first place (which itself produces distinctions between different waves of gentrifiers and thus the 'new middle classes'). Cosmopolitanism here leads to 'the development of a peculiar virtual and privatised landscape in which, despite its apparent "buzz", social interactions are limited, with very little possibility of accidental meetings – there are increasingly fewer "local" pubs or shops, for example' (Butler 2003: 2476).

Branded cosmopolitanism? Quarters, 'bubbles' and enclaves

Despite the apparent sense of cosmopolitanism in spaces such as the gentrified inner city, it might be that the cosmopolis is as much a space of containing diversity as one of appreciating and valuing difference. Indeed, these spaces are such that we may be seeing the development of a global geography of gentrified, seemingly cosmopolitan enclaves. Certainly, the respondents of Rofe's (2003: 2521) study identify the development of this curious geography in that, as one of the interviewees puts it, 'people from Glebe could go to these places and feel at home, that is what being a global member is all about, being comfortable in other places due to similar lifestyles – a frame of reference'. In fact it is not just the new middle classes that potentially move through these safe enclaves: it has been argued that the professional transnational elite also perform their globality in similar sanitised and homogeneous 'bubbles' (Szerszynski and Urry 2002: 469). Here we again see a spatialisation of cosmopolitanism in the distinctions that are often (erroneously) made between a mobile cosmopolitan elite able to move across space and disempowered locals trapped in place (Featherstone 2002). However, the erstwhile cosmopolitan practices of transnational elites are actually grounded and spatially concentrated in what Hannerz calls 'occidental cultural enclaves' (Hannerz 1996: 245). Again we see here how the apparent stateless and almost free-floating cosmopolite is actually grounded in quite specific sites and places wherein encounters with diversity and the development of a cosmopolitan attitude are seemingly contained and limited. The question then arises of whether such people or class formations are in fact cosmopolitan at all, in that they simply encounter and interact with other 'cosmopolitans' sharing the same cultural and aesthetic values.

The apparent desirability of gentrified and cosmopolitan spaces for redevelopment and regeneration is arguably resulting in what might be termed the serial reproduction of cosmopolitan spaces. Here we have the creation of spaces that share more in common with each other than with other spatial formations. In Nikolas Rose's essay on urban governance, in which he discusses practices of governance in contemporary urbanism (2000: 107):

> The city becomes not so much a complex of dangerous and compelling spaces of promises and gratifications, but a series of packaged zones of enjoyment, managed by an alliance of urban planners, entrepreneurs, local politicians and quasi-governmental 'regeneration' agencies.

Thus the production of cosmopolitan space has become part of political strategies for managing the city. Contemporary urban governance is focusing upon the production of commodified spaces of alterity and difference which, rather than generating new or challenging encounters, potentially result in a homogenisation and domestication of difference. Sandercock (2003) notes how in a number of contexts planning systems have failed to respond to the increasing cultural diversity of the city, partly because the values and norms of the dominant culture are embedded in its legislative frameworks and mundane operation. Consequently, we may be witnessing the serial reproduction of the same 'difference' across the globe, as Frank Mort (1998: 898–9) argues when he states that 'A hybrid version of Franco-Italianate café-bar society – serving the ubiquitous cappuccino and espresso – is now available in almost every metropolitan quarter in the world with claims to fashionability'. As Sandercock (2003: 125) suggests, the 'enclaving of the city' does not produce 'conditions conducive to an open and tolerant way of life', which contrasts strongly with her view that urban policy and planners should 'create the physical and discursive spaces . . . for renegotiations of collective identity' (2003: 151).

Contributing to, and closely aligned with, this serial reproduction of cosmopolitan spaces is the development of 'cultural quarters' (Bell and Jayne 2004). The development of these quarters and their branding as such has become a key strategy in contemporary urban governance, as part of an urban strategy of branding the city as global and thus making it desirable for inward investment, tourism and consumption. In this regard, ethnic and gay neighbourhoods and districts have come to be valued as assets in city promotion (Rushbrook 2002). Such spaces have been promoted and packaged as offering the visitor a genuine space of encounter with difference through consumption.

However, this is neither a universal process, nor has it taken the same form in each city. Within the same city we can witness different political strategies at a local level to develop such spaces. For instance, the work of Shaw *et al.* (2004) on two neighbourhoods in the East End of London – Brick Lane in Spitalfields close to the City of London, and Green Street in the London borough of Newham – reveals two different attempts at promoting multicultural districts. As they suggest, within promotion of this kind 'streets and neighbourhoods, whose very names once signified the poverty of marginalized communities, are repositioned to attract people with sophisticated, cosmopolitan tastes' (2004: 1983). These spaces are made attractive to outsiders of a particular social and class formation associated with affluence and the new middle class. However, there is a danger that this will result in their appropriation by an elite.

Local authorities and local entrepreneurs have been successful in making the Brick Lane area attractive through branding it as 'Banglatown'. But as Jacobs (1996: 100) shows, this branding trades on a 'racialised construct [of difference] tuned to multicultural consumerism'. Furthermore, this process of branding was perceived as divisive and exclusive by members of the local Bengali community and others. In comparison, the strategy adopted for Green Street by the London borough of Newham reflected this concern that promotion could lead to voyeurism by tourists, with one council official stating that they were 'reluctant to present it as a curiosity' (Shaw *et al.* 2004: 1995). Therefore, the Green Street policy has been to foster inclusivity and togetherness, rather than branding the area as the preserve of one ethnic group. Thus the strategy adopted here could be seen as more cosmopolitan because the space is not just promoted as the property of one ethnic group available for consumption, but rather the emphasis is on demonstrating that the area is receptive to multiple forms of ethnic and cultural difference.

Furthermore, Shaw *et al.* (2004) map out two possible scenarios for such spaces. The optimistic vision is that tourist revenues will create prosperity in ethnic minority businesses, and investment will make these neighbourhoods better places to live. However, the more pessimistic outlook suggests that the one-way traffic and attention of onlookers will become intrusive, disturbing people's everyday lives and reducing the quality of their local environment. In particular, the danger is that the signposting of difference will produce 'an anodyne and relatively homogenous culture of consumption . . . an isolated, tourism-oriented enclave [in] contrast to the poverty of the adjacent inner-city areas' (Shaw *et al.* 2004: 1997). Rising rents may force residents and businesses to leave the area, changing its distinctive identity and creating a bland and homogeneous district (see also Sandercock 2003). This example brings into sharp focus the tensions at the heart of contemporary cosmopolitan urbanism. On the one hand, there is the possibility that the quartering and branding of the city generates revenue and enables ethnic businesses to prosper. On the other hand, the branding of such spaces may increase their desirability, displacing poorer residents and paradoxically producing less socio-culturally diverse spaces, thus reducing the distinctiveness that made them attractive in the first place.

However, we must not ignore how ostensibly cosmopolitan quarters or districts of cities may be supportive of certain identities. Thus areas that are not the property of one single ethnic or socio-cultural group may be attractive to those whose identities may transcend these labels. For example, this is suggested in Hiebert's (2002) study of Vancouver. For many of the 400,000 immigrants who settled in Vancouver during the 1990s, language difficulties and cultural differences may make everyday survival and getting by in the city highly stressful, leading them to become dependent on transnational networks for support. As a result 'cosmopolitanism, in an active sense, occurs rarely given the struggle for survival and communication barriers' (Hiebert 2002: 215). Hiebert's study reveals the difficulties for immigrants of obtaining work within Vancouver and how instances of everyday racism make incomers more reliant on transnational networks. This racism was not simply expressed as the racism of the host community versus the outsider. Hiebert (2002: 218, emphasis in original) argues that 'one of the most consistent findings of both focus groups and family interviews was that immigrants reserve their most bitter criticism for members of *other minority groups*'. However, Hiebert also found that many immigrants displayed cosmopolitan tendencies by seeking to live in multicultural neighbourhoods rather than enclaves predominantly associated with their own ethnic community, out of a desire to better integrate into Canadian society, or to escape the confines of their own ethnic group. In summary, he argues that (2002: 223):

> as minorities increasingly become hosts, and with all of the power relations that such a turn of events implies, they take on the same responsibilities to extend fairness and hospitality to newcomers – in effect, to be cosmopolitan. It is hoped that they will exercise these responsibilities more conscientiously than the dominant group has done.

What this study highlights is not only the complexity through which cosmopolitanism is spatialised and articulated in the city, but also the need to ground studies of cosmopolitanism in specific spatial and temporal contexts. Sensitivity to the complexity of the context of its formation is necessary for an understanding of cosmopolitanism and cosmopolitan urbanism. Hiebert (2002: 210) sets out a series of questions about what he terms 'the lived experience of . . . cosmopolitanism' which are useful in framing this complex and contextual relationship between cosmopolitanism and the city: how does cosmopolitanism arise at the micro-geographical scale of the neighbourhood? Who participates (and who does not) in cosmopolitan interaction? Finally, what does it mean to grow up in a cosmopolitan setting?

Cosmopolitan urbanism: section and chapter outlines

This volume is structured into three parts. The chapters and case studies contained in each overlap and offer different conceptions of cosmopolitanism and examples of how these notions are operationalised in different urban contexts. However, they can be grouped around three approaches to cosmopolitan urbanism: envisaging cosmopolitan urbanism; consuming the cosmopolitan city; and producing the cosmopolitan city. While these categories are not mutually exclusive, they each offer a certain way of approaching cosmopolitan urbanism and how it is applied in different contexts. This section thus provides a brief outline to each part of the book together with an overview of the chapters they contain.

Part I – Envisaging cosmopolitan urbanism

If, as this introduction has suggested, 'cosmopolitanism' may be seen as an empty signifier which is filled with different content in different situated cultural worlds, then central to this book is the examination of how it is filled in different urban contexts. The first section thus features three chapters which focus on conceptualisations of the term in the context of various modern urban settings. At an abstract level cosmopolitanism can be conceived of as a philosophy of global citizenship or a specific attitude towards difference, but how are these abstract notions grounded and operationalised? And what are the implications for cosmopolitanism of considering it in the light of differing urban experiences? Sandercock (2003: 2) is one of the few authors who attempt to offer a definition of what should emerge if abstract notions of cosmopolitanism engage with the urban when she talks of the ideal of 'cosmopolis' as 'a city . . . in which there is genuine acceptance of, connection with, and respect and space for the cultural other, and . . . the possibility of a togetherness in difference'.

However, cosmopolitanism is a highly contested concept. As Robbins (1998: 12) suggests, the term 'points . . . to a domain of contested politics'. In this light Beck (2002: 29) warns of the danger of the 'cosmopolitan fallacy' – the idea that we can find a cosmopolitan state of existence:

> Even the most positive development imaginable, an opening of cultural horizons and a growing sensitivity to other unfamiliar, legitimate geographies of living and coexistence, need not necessarily stimulate a feeling of cosmopolitan responsibility.

Beck thus posits an ongoing process of 'cosmopolitanization' as being 'about a dialectics of conflict: cosmopolitanization *and its enemies*' (2002: 29, emphasis in original). These enemies include nationalism, global capitalism and democratic authoritarianism. In other words, Beck (2002) is suggesting that the potential for becoming cosmopolitan is constantly in conflict with other tendencies in contemporary society, such as continued and even strengthened allegiances to national and other identities or the rise in the power of the state and a diminished potential for achieving democratic consensus. To understand processes of 'cosmopolitanization' we must examine how they are occurring in different situated contexts. Doing this in the contemporary urban setting implies acknowledging that, once grounded, 'cosmopolitanism' is a highly contested concept, intersecting with the multiply contested politics of class, gender and sexuality, race and ethnicity, and power in the city. In addition, cosmopolitanism and cities are not just about the local, but develop and are governed within complex national and global–local contexts.

These concerns are taken up by Sandercock's chapter, which addresses cosmopolitan urbanism within the context of globalisation and migration. Due to these processes cities are becoming more significant as sites of hybridity, difference and diversity. Sandercock questions how city dwellers can learn to live with each other. She argues that traditional ways of conceptualising cosmopolitanism privilege transnational elites and are inadequate in dealing with the new diversity that the individual faces. Consequently, Sandercock examines lived experience in cities, outlining the 'cosmopolitan project' as an intercultural and political project which must continue to recognise the inevitability of encountering difference and the potential for conflict this may produce. Sandercock's vision is one of 'pragmatic urbanity' where intercultural differences and conflicts are resolved through mundane everyday practices. She points to the failure of urban planners to adequately produce spaces in the city which function as sites of 'interdependence and habitual engagement', and draws upon Amin's (2002) concept of 'micropublics' in setting out an agenda whereby cosmopolitan practices and discourses can be realised (see also Germain and Radice, this volume). She provocatively suggests we subscribe to a wider faith in democratic and consensual processes of conflict resolution through spaces which can accommodate intercultural dialogue.

Contrasting with Sandercock's vision, Bridge examines the link between cosmopolitan knowledge and circuits of cultural capital in gentrification in Bristol and London (UK) and Sydney (Australia), highlighting how these processes fail to produce spaces accommodating intercultural dialogue. This chapter explores how professional groups in these cities derive cultural capital through localised residential strategies and through choices of schooling for their children. Such social groups appear at ease in the world, possessing the appropriate competencies, attitudes and transnational connections to smoothly navigate through encounters with cultural difference. However, Bridge argues that their cosmopolitanism is derived from the selective use of cultural capital, implying a paradox between simultaneously embracing certain forms of difference whilst devalorising others. This selection process is spatialised by the creation of gentrified 'cosmopolitan' enclaves. The lifestyle and consumption patterns associated with gentrification and the discursive community it produces serves to 'lock in' professional knowledge-elites within narrowly defined neighbourhoods distinguished by specific ethnic and class status and the deployment of taste or aesthetic displays of housing choice and school preferences. The resultant metropolitan habitus naturalises distinct class formations through the internal transmission of cultural knowledge, disposition and taste. Paradoxically, therefore, supposedly cosmopolitan individuals become isolated from the daily disturbances that would otherwise occur through encountering difference, denying opportunities for intercultural dialogue.

Taking these points further, Iveson's chapter is concerned that the city's position as a place of refuge for strangers is threatened by an 'enclave consciousness'. Whereas urban life is very much about 'life amongst strangers', Iveson contrasts this notion with an escalating fear of strangers. This fear of the 'Other' is spatialised through planning interventions which create gated communities or extend regulation and surveillance. Iveson is critical of the political paranoia surrounding attempts to legally define and limit rights to exist in a certain place, which are afforded to those we know and not to those whom we fear. Against this climate of anxiety, transnational migration is escalating encounters with strangers and challenges the notion of hospitality, placing social and spatial limits to hospitality in terms of our obligations to the stranger. The ability to welcome is seen as a cosmopolitan virtue, privileging particular class and ethnic groups who possess the appropriate skills and attitude to provide the welcome. In doing so the 'welcomers' are able to outline the terms of hospitality and what difference should be accommodated. This requires a point of adjustment on behalf of the stranger with an obligation to 'fit in'. Iveson, however, argues that the complexity and diversity of modern urban life is producing universal estrangement. In effect we are all partial strangers, thus raising the challenging question of how all of us deal with this condition. Iveson's argument, therefore, challenges the barrier between host and stranger, moving from a dichotomous relationship into a more fluid dialogue between self and other. Finally, Iveson argues that a cosmopolitan state of being rests on the need for openness and reasonableness where differences are worked out through everyday practice.

Part II – Consuming the cosmopolitan city: materialities and practices

Through gentrification and the branding of areas within the city as 'cosmopolitan', cosmopolitan urbanism has an intimate link with the symbolic and material territorialisation of difference. Indeed, cosmopolitanism is often associated with areas of cities which are explicitly branded and promoted as 'cosmopolitan'. In this sense cosmopolitanism is associated with particular spaces in the city, especially where particular forms of consumption are engaged with and performed. Thus we see a spatialisation of particular forms of 'cosmopolitan consumption', involving new arrangements of people, objects and performances by consumers who can conform to and engage with this particular vision of a 'cosmopolitan lifestyle'.

There is a temptation here to dismiss these consumption spaces as superficial, commodified expressions of cosmopolitanism that exist because of contemporary trends in maximising economic value in gentrifying city centres. Certainly they may become cosmopolitan 'enclaves' in which those who can perform the 'right' kind of cultural capital can indulge in a superficial encounter with difference and the 'Other' through shopping, the café-bar culture or 'ethnic' restaurants without actually knowing the 'Other' in any way. The planning of such spaces to support 'cosmopolitan' consumption practices can also, intentionally or unintentionally, act to homogenise them and exclude difference (Raco 2003). The mundane activities of urban authorities that support a certain style of regeneration, or which monitor and regulate certain ('public') spaces so as to make them 'safe' for consumers, or of businesses which sell products that appeal to a certain 'cosmopolitan aesthetic', can all lead to unintended exclusions of difference. This 'domestication by cappuccino', in which public space is improved and 'cleansed' to support consumption, leading to the systematic exclusion of some groups, has been commented on in a variety of urban contexts (Zukin 1988; Miles 1998; Atkinson 2003). What these accounts highlight is that such developments in the contemporary city relate in complex ways to a spatialisation of difference in which certain groups and areas within cities are labelled as 'cosmopolitan' or 'non-cosmopolitan' on the basis of consumption. Further, this labelling is related to processes of defining and naturalising 'acceptable' and 'unacceptable' forms of difference in these areas.

However, it is important not to read off the nature of social relations from the outward appearances of such spaces, and the four chapters in this section challenge any simplistic reading of cosmopolitan consumption spaces. First, Latham examines the processes through which cosmopolitanism comes into being by exploring how urban lives are exposed to a diverse range of non-local cultural influences. However, rather than producing a homogeneous 'one-world' culture, Latham argues that cosmopolitan influences create heterogeneous local experiences and material cultures through our social encounters and use of material objects. He draws a distinction between traditional understandings of the cosmopolitan self and those people whose cosmopolitan experience is more grounded in local contexts. Notionally, cosmopolitans include people who are 'at home in the world' and possess the appropriate competences, attitude and knowledge to engage with difference. However, Latham discusses people who are 'of the world', who possess an openness to difference, but where cosmopolitanism is grounded in everyday urban experiences rather than transnational encounters. Through examining the lives of residents negotiating new 'cosmopolitan' consumption spaces in Auckland (New Zealand), Latham reveals the hybridity produced from the clash between transnational trends in cosmopolitan consumption and local cultures keen to celebrate exotic difference. In particular, he examines the impact on material culture, social practices and the use of spaces of hospitality within the city. Latham argues that this hybridity extends possibilities or opportunities to further a sense of national identity. In his words, material and aesthetic practices 'become entangled in a whole new mesh of practice and meaning' in such consumption spaces.

Germain and Radice's chapter examines the tension created through attempts by the local state to plan for a cosmopolitan city. Montréal (Canada) provides an interesting context where national and local identities are keenly contested. They pose the question of where cosmopolitanism is constituted by examining aspects of sociability in consumption spaces, public spaces and communities. In particular, they ask whether cosmopolitanism can be achieved through state-sponsored city-wide collective projects. They examine the tension between 'official policies' to mould new forms of citizenship and the everyday practices of encounters between diverse cultural groups. This leads to a wider question of whether the local state has the capacity to produce a more cosmopolitan city, raising concerns that cosmopolitanism cannot be planned for and is perhaps best arrived at through quotidian experiences. Instead they propose Amin's (2002) concept of micro-publics (and see also Sandercock, this volume) as a way of resolving intercultural conflicts and differences, implying that cosmopolitanism can be produced by the citizenry itself. In this sense it is particular urban spaces, including consumption spaces, that foster pragmatic encounters between diverse social groups.

Brown's chapter examines the intersection of cosmopolitanism and sexual/class difference, focusing on Spitalfields, London (UK), which is home to many Bengalis, but also accommodates sexual dissidence. Regeneration strategies have promoted Spitalfields as a cosmopolitan enclave, but the rebranding of the area as 'Banglatown' emphasises the area's ethnic rather than sexual diversity. Despite the presence of a significant number of gay and queer sites, Spitalfields is not necessarily marked out as a gay space, but functions as a zone of cosmopolitan consumption. However, this attempt to plan cosmopolitan (consumption) space has produced a form of cosmopolitan urbanism which enables the strategic deployment of cultural capital allowing both gay and heterosexual men, of diverse ethnic and class backgrounds, to seek out erotic encounters through everyday contact and engagement. In this context, therefore, cosmopolitanism is used as a form of 'camouflage' by marginalised non-heterosexual men to explore their desires and identity, without having to subscribe to a mainstream 'gay' or 'queer' identity. Brown sees this development as a positive process in the development of the city, showing how diverse groups can experience each other and coexist. Although the rebranding of Spitalfields appears to be another superficial exercise in place marketing and 'cosmopolitan consumption', it has perhaps unintentionally created a space in which gay and queer men also encounter and negotiate ethnic and class differences.

In Singapore, the national government is attempting to produce a more cosmopolitan society, to create a place where cosmopolitans can feel 'at home', whilst at the same time 'cosmopolising' Singaporeans by encouraging the spread of cosmopolitan virtues and values. The ultimate aim of the state, however, is to build a stronger sense of national unity, by fostering feelings of belonging to their birthplace within both Singapore's transnational elites and its 'rooted' local communities. In Tan and Yeoh's chapter, therefore, we can see both the spatialisation of cosmopolitanism, and how the state can define and attempt to naturalise 'acceptable' and 'unacceptable' forms of difference in certain areas of the city. Government discourse, in particular, draws a distinction between 'cosmopolitans' – people who are 'rootless', mobile and educated – and the 'heartlanders' – those who are 'rooted' and immobile. Tan and Yeoh's analysis focuses on how this discourse is reflected in the narratives of popular fiction, revealing a complex set of relationships and conflicts between cosmopolitan and non-cosmopolitan subjectivities. In Singapore, attempts to geographically transform the 'heartland' into cosmopolitan (consumption) space produces an uneasy mix of the modern with deep-seated cultural heritage, highlighting tensions between cosmopolitanism and nationalism, the global and the local, and Western and Eastern cultures.

Part III – Producing the cosmopolitan city: cultural policy and intervention

The third section of this book considers the lesser studied processes which we refer to as the production of cosmopolitan space. This does not employ a simplistic distinction between production and consumption, but the focus here is on the production of cosmopolitan spaces in cities as a part of broader strategies of urban development. The emphasis is thus on how key actors – particularly the local state and policy communities – are constructing, deploying and promoting notions of the 'cosmopolitan city'. Here we examine the grounding of cosmopolitanism in the city through its interweaving into urban policy. This happens in the contemporary city in a number of ways. The local state, for example, may construct and brand the city, or commodified spaces within it, as 'cosmopolitan' to attract investment, businesses or tourism. Cosmopolitanism is also used explicitly in local policies directed at managing multiethnic or 'multicultural' areas within cities. In this context, however, cosmopolitan urbanism begins to intersect with neo-liberal forms or 'entrepreneurial' forms of urban governance (Harvey 1989), as it is particularly within such approaches to urban regeneration that notions of the 'cosmopolitan city' are defined. Neo-liberal or entrepreneurial urbanisms commonly ground notions of cosmopolitanism in the planning, remodelling and reimaging of the 'post-industrial' city, using it as a resource in competitive economic development strategies.

However, again these processes involve an explicit labelling of cities or parts of the city as 'cosmopolitan' and, by implication, other spaces as 'non-cosmopolitan'. Gentrification, 'loft living', the remodelling of the city centre towards consumption, and 'entrepreneurial' urban governance have all been linked to the exclusion of unwanted forms of difference, particularly of those who will not or cannot 'fit into' or participate in the dominant style of, or vision for, urban development (Brenner and Theodore 2002; Macleod 2002; Smith 2002). This links the grounding of cosmopolitanism in the city to the nature of political power and governance and the impact on diversity and social justice.

The production of cosmopolitanism in contemporary urban development strategies thus illustrates its paradoxical nature. The construction of cosmopolitan spaces implies that there must be some form of 'Other' space, of 'non-cosmopolitan' difference, against which it is constructed. Is there then a paradoxical displacement of other forms of 'disruptive' difference which need to be 'purified' from 'cosmopolitan' spaces as they are upgraded for economic ends? The notion of the 'revanchist city' in particular stresses the increasingly programmatic nature of city governance which focuses on the poor and dispossessed as part of a disciplinary and authoritarian stance towards the city (Smith 1996). More subtly Flusty's (2001) notion of 'interdictory space' points to how apparently public spaces are actually regulated and subject to surveillance. Thus, as Keohane (2002: 42) suggests, there is a danger that these new and apparently 'cosmopolitan' city spaces become 'homogenized spaces . . . from which possibility of encounter with other forms of life has been all but eliminated'.

Again, it is necessary to ground such questionings of cosmopolitan urbanism in the realities of life in such spaces, particularly given the more subtle readings of such spaces indicated in Part II above, and to be aware of the continued importance of national context for cosmopolitanism. In the first chapter of this part of the book Bodaar examines the notion of the multicultural state through a case study of community development in Amsterdam's Bijlmermeer district (the Netherlands), a densely populated, planned suburb of high-rise estates accommodating a diverse range of mainly working-class non-Dutch immigrants. After decades of decline, the local state is attempting to regenerate this suburb, rebranding the district as a 'cosmopolitan' space with the intention of marketing Bijlmermeer's exotic ethnic difference to attract investment and middle-class residents. Local residents, however, have resisted this attempt to produce a sanitised or superficial cosmopolitanism. Bodaar connects this local struggle to a national political debate concerning how the Dutch state has dealt with multiculturalism in an attempt to forge a greater sense of national unity, through the regulation and legitimation of certain differences but not others. This has become spatialised through planning interventions to integrate diverse communities by engineering 'cosmopolitan' city districts, which involve an implicit labelling of local communities as cosmopolitan or non-cosmopolitan. Bodaar, however, reveals a paradox, wherein cosmopolitanism, in the sense of openness to difference, is perhaps more likely to be found within those spaces ostensibly labelled as non-cosmopolitan.

The key focus of Haylett's chapter is her attempt to relate the often neglected significance of class culture to the analysis of cosmopolitanism. She focuses on the Family Pathfinders scheme in Houston (Texas, USA), a project designed to assist people to find work. The majority of those taking part are welfare-dependent African-American women with children. Pathfinders involves middle-class volunteers lending support to these families by acting as mentors, to facilitate an exchange of aspirational values. It is hoped that individuals will build up their self-esteem and eventually find work through an almost therapeutic process of self-empowerment. Haylett sees the scheme as an attempt to forward a nationally defined notion of active citizenship through a stretching of openness across class boundaries to connect dependent and independent social groups. Haylett's analysis draws a distinction between the 'cosmopolitan self' and working-class 'Others', focusing on how national welfare regimes play a central role in furthering social inequalities through reinforcing class cultural differences. The cosmopolitan is marked out by their identification with the national political project framed by a neo-liberal welfare agenda prioritising economic goals, Christian values, and a particular attitude towards racial and gender differences. In this process, however, the working-class, non-cosmopolitan 'Other' is forced into a position of dependency through a withdrawal of state benefits. Haylett questions whether a 'cosmopolitan city' can ever exist whilst class difference survives.

Chan's chapter explores issues of race and ethnicity in the branding of Birmingham (UK) as a cosmopolitan city by the local state. Central to this strategy is the promotion of a model of active citizenship, which fuses neo-liberal notions of self-help and entrepreneurialism with Birmingham's identity as an ethnically diverse city. By pushing forward a model of interculturalism, the city aims to produce a more cosmopolitan and welcoming environment for 'Others'. Focusing on the promotion of Birmingham's Chinese Quarter, Chan critically examines the notion of ethnic minorities as a cultural and economic asset for the city. The 'cosmopolitan' nature of Birmingham is seen as a resource for developing the city, but as Chan demonstrates it is a superficial view of cosmopolitanism in which only certain forms of 'ethnic' difference are valued. This analysis of the city's 'welcome' to 'Others' reveals a process fraught with exclusion and paradox, highlighting the limits to Birmingham's hospitality.

In the final chapter Binnie and Skeggs examine the production and branding of Manchester's Gay Village (UK) as a space of sexualised cosmopolitan consumption. Their argument first examines the capacity of business to find economic opportunities through accentuated spatial divisions of consumption which involve explicit marketing strategies drawing upon particular cultural forms. In this context, despite the 'authentic' roots behind the development of the Gay Village, the area is becoming commodified. To present the Village to a mainstream target group the area is openly branded as a cosmopolitan space, which draws upon its reputation as a zone of tolerance and openness. However, this requires a recasting of the role of sexualised 'Others' who are presented as non-threatening and safe. Binnie and Skeggs examine in particular how privileged non-gays are able to deploy cultural capital to navigate through the Village and negotiate the difference they encounter, opening up a part of the city to them for new experiences and opportunities. The marketing of sexual dissidence, however, is not necessarily an all-embracing process, as it values certain forms of male homosexual culture whilst simultaneously marginalising 'Other' gay individuals. The cosmopolitan branding of space has extended beyond the Village to encompass other parts of the city centre, serially reproducing a certain look or feel within the urban environment, but not necessarily affording similar opportunities for gays and lesbians.

As this introduction has argued, cosmopolitanism as a concept is multiply defined and variously contested. The aim of this volume is to attempt to ground the concept in the urban to examine how it is constructed, deployed and contested in specific contexts. The issues identified above are thus explored in more depth in the set of chapters which follow (and which have been outlined above). Each of these chapters addresses a particular take on cosmopolitan urbanism, through considering specific case studies of the operationalising of the term in particular urban contexts, within three sections on envisaging cosmopolitan urbanism, consuming the cosmopolitan city and producing the cosmopolitan city. Finally, the conclusion to the book reflects on the key themes which these chapters address before making critical suggestions for further research into cosmopolitan urbanism.

References

Amin, A. (2002) 'Ethnicity and the multicultural city: living with diversity', *Environment and Planning A*, 34: 959–80.

Appiah, K. A. (1998) 'Cosmopolitan patriots', in P. Cheah and B. Robbins (eds), *Cosmopolitics: Thinking and Feeling Beyond the Nation*, Minneapolis, MN: University of Minnesota Press.

Atkinson, R. (2003) 'Introduction: misunderstood saviour or vengeful wrecker? The many meanings and problems of gentrification', *Urban Studies*, 40: 2343–50.

Beck, U. (2002) 'The cosmopolitan society and its enemies', *Theory, Culture and Society*, 19: 17–44.

Beck, U. (2004) 'Cosmopolitical realism: on the distinction between cosmopolitanism in philosophy and the social sciences', *Global Networks*, 4: 131–56.

Bell, D. and Jayne, M. (2004) *City of Quarters: Urban Villages in the Contemporary City*, Aldershot: Ashgate.

Benedictus, L. (2005) 'London: the world in one city: a special celebration of the most cosmopolitan place on earth', *Guardian* G2, 21 February 2005, pp. 1–7.

Bennetto, J. (2005) 'London's cosmopolitan criminals targeted', *Independent*, 5 February 2005, p. 15.

Brenner, N. and Theodore, N. (2002) 'Preface: from the "new localism" to the spaces of neoliberalism', *Antipode*, 34: 341–7.

Bridge, G. (2001) 'Estate agents as interpreters of economic and cultural capital: the gentrification premium in the Sydney housing market', *International Journal of Urban and Regional Research*, 25: 87–101.

Butler, T. (2003) 'Living in the bubble: gentrification and its "Others" in North London', *Urban Studies*, 40: 2469–86.

Calhoun, C. (2002) 'The class consciousness of frequent travellers: towards a critique of actually existing cosmopolitanism', in S. Vertovec and R. Cohen (eds), *Conceiving Cosmopolitanism: Theory, Context, and Practice*, Oxford: Oxford University Press.

Cheah, P. and Robbins, K. (eds) (1998) *Cosmopolitics: Thinking and Feeling Beyond the Nation*, Minneapolis, MN: University of Minnesota Press.

Demos (2003) 'Ethnic diversity and gay people are key indicators of cities' creative potential, according to new UK creativity index', Demos media release, 26 May 2003, London. Available at www.demos.co.uk/media/_page266.apx.

Featherstone, M. (2002) 'Cosmopolis: an introduction', *Theory, Culture and Society*, 19: 1–16.

Fine, R. (2003a) 'Kant's theory of cosmopolitanism and Hegel's critique', *Philosophy and Social Criticism*, 29: 609–30.

Fine, R. (2003b) 'Taking the "ism" out of cosmopolitanism: an essay in reconstruction', *European Journal of Social Theory*, 6: 451–70.

Florida, R. (2002) *The Rise of the Creative Class*, New York: Basic Books.

Flusty, S. (2001) 'The banality of interdiction: surveillance, control and the displacement of diversity', *International Journal of Urban and Regional Research*, 25: 658–64.

Hannerz, U. (1996) *Transnational Connections: Culture, People, Places*, London: Routledge.

Harvey, D. (1989) 'From managerialism to entrepreneurialism: the transformation of urban governance in late capitalism', *Geografiska Annaler B*, 71: 3–17.

Hiebert, D. (2002) 'Cosmopolitanism at the local level: the development of transnational neighbourhoods', in S. Vertovec and R. Cohen (eds), *Conceiving Cosmopolitanism: Theory, Context, and Practice*, Oxford: Oxford University Press.

Isin, E. and Wood, P. (1999) *Citizenship and Identity*, London: Sage.

Jacobs, J. M. (1996) *Edge of Empire: Postcolonialism and the City*, Routledge: London.

Kennedy, P. (2004) 'Making global society: friendship networks among transnational professionals in the building design industry', *Global Networks*, 4: 157–79.

Keohane, K. (2002) 'The revitalization of the city and the demise of Joyce's utopian modern subject', *Theory, Culture and Society*, 19: 29–50.

Kiely, R. (2004) 'What difference does difference make? Reflections on neo-conservatism as a liberal cosmopolitan project', *Contemporary Politics*, 10: 185–202.

Lamont, M. and Aksartova, S. (2002) 'Ordinary cosmopolitanisms: strategies for bridging racial boundaries among working class men', *Theory, Culture and Society*, 19: 1–26.

Latham, A. (2003) 'Urbanity, lifestyle and the politics of the new economy', *Urban Studies*, 40: 1699–724.

Ley, D. (2004) 'Transnational spaces and everyday lives', *Transactions of the Institute of British Geographers*, 29: 151–64.

Macleod, G. (2002) 'From urban entrepreneurialism to a "revanchist city"? On the spatial injustices of Glasgow's renaissance', *Antipode*, 34: 602–24.

May, J. (1996) 'Globalization and the politics of place: place and identity in an inner London neighbourhood', *Transactions of the Institute of British Geographers*, 21: 194–215.

Miles, S. (1998) 'The consuming paradox: a new research agenda for urban consumption', *Urban Studies*, 5–6: 1001–8.

Mort, F. (1998) 'Cityscapes: consumption, masculinities, and the mapping of London since 1950', *Urban Studies*, 35: 889–907.

Nava, M. (2002) 'Cosmopolitan modernity: everyday imaginaries and the register of difference', *Theory, Culture and Society*, 19: 81–99.

Pécoud, A. (2004) 'Entrepreneurship and identity: cosmopolitanism and cultural competences among German–Turkish businesspeople in Berlin', *Journal of Ethnic and Migration Studies*, 30: 3–20.

Raco, M. (2003) 'Remaking place and securitising space: urban regeneration and the strategies, tactics and policies of policing in the UK', *Urban Studies*, 40: 1869–87.

Reiss, H. (ed.) (1991) *Kant's Political Writings*, Cambridge: Cambridge University Press.

Robbins, B. (1998) 'Introduction part 1: actually existing cosmopolitanism', in B. Robbins and P. Cheah (eds), *Cosmopolitics: Thinking and Feeling Beyond the Nation*, Minneapolis, MN: University of Minnesota Press.

Rofe, M. W. (2003) '"I want to be global": theorising the gentrifying class as an emergent elite global community', *Urban Studies*, 40: 2511–26.

Rose, N. (2000) 'Governing cities, governing citizens', in E. Isin (ed.), *Democracy, Citizenship and the Global City*, London: Routledge.

Rushbrook, D. (2002) 'Cities, queer space and the cosmopolitan tourist', *GLQ: A Journal of Lesbian and Gay Studies*, 8: 183–206.

Sandercock, L. (2003) *Cosmopolis II. Mongrel Cities of the 21st Century*, London: Continuum.

Schein, L. (1999) 'Of cargo and satellites: imagined cosmopolitanism', *Postcolonial Studies*, 2: 345–75.

Shaw, S., Bagwell, S. and Karmowska, J. (2004) 'Ethnoscapes as spectacle: reimagining multicultural districts as new destinations for leisure and consumption', *Urban Studies*, 41: 1983–2000.

Sklair, L. (2001) *The Transnational Capitalist Class*, Oxford: Blackwell.

Smith, M. P. and Guarnizo, E. L. (eds) (1998) *Transnationalism From Below. Comparative Urban and Community Research*, vol. 6, New Brunswick: Transaction Publishers.

Smith, N. (1996) *The New Urban Frontier: Gentrification and the Revanchist City*, London: Routledge.

Smith, N. (2002) 'New globalism, new urbanism: gentrification as global urban strategy', *Antipode*, 34: 427–50.

Stevenson, N. (2002) 'Cosmopolitanism, multiculturalism and citizenship', *Sociological Research Online*, 7. Available at www.socresonline.org.uk/7/l/stevenson.html.

Szerszynski, B. and Urry, J. (2002) 'Cultures of consumption', *Sociological Review*, 50: 461–81.

Turner, B. (2000) 'Cosmopolitan virtue: loyalty and the city', in E. Isin (ed.), *Democracy, Citizenship and the Global City*, London: Routledge.

Werbner, P. (1999) 'Global pathways. Working class cosmopolitans and the creation of transnational ethnic worlds', *Social Anthropology*, 7: 17–35.
Yeoh, B. (2004) 'Cosmopolitanism and its exclusions in Singapore', *Urban Studies*, 41: 2431–45.
Zukin, S. (1988) *Loft Living: Culture and Capital in Urban Change*, London: Radius.

PART I

Envisaging cosmopolitan urbanism

Cosmopolitan urbanism: a love song to our mongrel cities

Leonie Sandercock

Introduction

> If the Satanic Verses is anything, it is a migrant's eye view of the world. It is written from the very experience of uprooting, disjuncture and metamorphosis (slow or rapid, painful or pleasurable) that is the migrant condition, and from which, I believe, can be derived a metaphor for all humanity. Standing at the center of the novel is a group of characters most of whom are British Muslims . . . struggling with just the sort of great problems that have arisen to surround the book, problems of hybridization and ghettoization, of reconciling the old and the new. Those who oppose the novel most vociferously today are of the opinion that intermingling with a different culture will inevitably weaken and ruin their own. I am of the opposite opinion. The Satanic Verses celebrates hybridity, impurity, intermingling, the transformation that comes of new and unexpected combinations of human beings, cultures, ideas, politics, movies, songs. It rejoices in mongrelization and fears the absolutism of the Pure. Melange, hotchpotch, a bit of this and a bit of that is how newness enters the world. It is the great possibility that mass migration gives the world . . . The Satanic Verses is for change-by-fusion, change-by-conjoining. It is a love song to our mongrel selves.
>
> (Rushdie 1992: 394)

In this defence of his controversial novel, Salmon Rushdie staked out some of the territory that I want to cover in this chapter. I want to use the metaphor of the mongrel city to characterize an emerging urban condition in which difference, otherness, multiplicity, heterogeneity, diversity and plurality prevail. For some this is to be feared, signifying the decline of civilization as we know it in the West. For others it is to be celebrated as a great possibility: the possibility of living alongside others who are different, learning from them, creating new worlds with them, instead of fearing them. My recent project has been to provide a better understanding of the emergence of *cities of difference* in the context of globalization and other related social forces; and to reflect on the challenges which these mongrel cities present in the twenty-first century to the city-building professions (architects, planners and urban designers, landscape architects, engineers), to city dwellers, and to conventional notions of citizenship (Sandercock 2003). My central question is how can 'we', (all of us), in all of our differences, be 'at home' in the increasingly multicultural and multiethnic cities of the twenty-first century? Or, as James Donald (1999) puts it more vigorously, how can we stroppy strangers live together in these (mongrel) cities without doing each other too much violence? That seems to me to be the central and defining question for a cosmopolitan urbanism.

By the late twentieth century, cosmopolitanism as a concept/world view was regarded with considerable disapproval by a variety of respected theorists. David Harvey (2000) has critiqued the Kantian origins of cosmopolitanism as 'nothing short of an intellectual and political embarrassment', based largely on Kant's egregious racism. Peter Van der Veer (2002) dismisses cosmopolitanism for its complicity in the centuries-long Western colonial project. Craig Calhoun (2002), in a devastating turn of phrase, portrays cosmopolitanism as the preferred ethical orientation of those privileged to inhabit the frequent-traveller lounges. Yet, in spite of this bad rap, there has been a resurgence of cosmopolitan theorizing since the mid-1990s. Hollinger has given persuasive reasons for the emergence at this historical moment of formulations of a new cosmopolitanism. Among the historical circumstances that have most obviously helped to call forth this movement, he argues, are the dead ends reached by identity politics within the USA, the destruction caused by ethno-religious nationalism in the wake of the end of the Cold War, and the challenges to provincial orientations presented by globalization (Hollinger 2002: 228). Various other authors and volumes have sought to make a case for a 'new cosmopolitanism' (Nussbaum 1996; Cohen and Nussbaum 1996; Brennan 1997; Cheah and Robbins 1998; Beck 1999; Falk 2000; Zachary 2000; Hollinger 2002; Vertovec and Cohen 2002). Two of these authors have gone so far as to issue 'cosmopolitan manifestos' (Nussbaum 1996; Beck 1999). While there is no shared political philosophy among these new cosmopolitan theorists, they do share a preoccupation with such global issues as international peace and governance, the state of the environment, social development and human rights abuses, and a desire to stimulate an overall 'process of world thinking' (Vertovec and Cohen 2002: 21).

As an urbanist with an interest, beyond theory, in the actual conditions of existence in the world's cities, and in practical and policy questions around managing our peaceful coexistence in shared spaces, I bring a different spin to the new discourse on cosmopolitanism. Along with the geographers who conceived this volume of essays, I seek to harness cosmopolitan thinking to the actual spaces of cities, as sites of meaning making (of belonging), and of a located politics. In this chapter I want to argue for a *cosmopolitan urbanism* as a normative project that is a necessary response to the empirical reality of multicultural cities. Such a project has at least two dimensions: a social imaginary of living together in difference, and a political philosophy capable of overcoming the weaknesses of twentieth-century multiculturalism. In Part 1, I discuss three sociological imaginings and accounts of how we might live together in all of our differences. Through these different imaginings I explore what it means to be 'at home' in an increasingly globalized world; what a sense of belonging might be based on in a multicultural society; and how to encourage more intercultural encounters, exchanges, and solidarity. I take seriously Calhoun's argument that not only tolerance but also solidarity is required for people to live together and join in democratic self-governance (Calhoun 2002: 108). Part 2 begins with a brief critique of the twentieth-century multicultural project, which has sometimes been mistakenly identified with a cosmopolitan urbanism. It then proceeds to outline a twenty-first-century *intercultural* project as a more truly cosmopolitan project grounded in political community and agonistic democracy rather than ethno- (or any other sub-)cultural identity as a basis for a sense of belonging in mongrel cities.

Part 1: How might we live together in all of our differences? Three imaginings

Richard Sennett: togetherness in difference

In *Flesh and Stone* (1994: 358) Sennett laments that the apparent diversity of Greenwich Village in New York is actually only the diversity of the gaze, rather than a scene of discourse and interaction. He worries that the multiple cultures that inhabit the city are not fused into common purposes and wonders whether 'difference inevitably provokes mutual withdrawal'. He assumes that if the latter is true, then 'a multicultural city cannot have a common civic culture' (Sennett 1994: 358). For Sennett, Greenwich Village poses a particular question of how a diverse civic culture might become something people feel in their bones. He deplores the ethnic separatism of old multicultural New York and not only looks but longs for evidence of citizens' understanding that they share a common destiny. This becomes a hauntingly reiterated question: nothing less than a moral challenge, the challenge of living together not simply in tolerant indifference to each other, but in active engagement. For Sennett, then, there is a normative imperative in the multicultural city to engage in meaningful intercultural interaction.

Why does Sennett assume that sharing a common destiny in the city necessitates more than a willingness to live with difference in the manner of respectful distance? Why should it demand active engagement? He does not address these questions, nor does he ask what it would take, sociologically and institutionally, to make such intercultural dialogue and exchange possible, or more likely to happen. But more recently other authors have begun to ask, and give tentative answers to, these very questions (Donald 1999; Parekh 2000; Amin 2002; Sandercock 2003). In terms of political philosophy, one might answer that in multicultural societies composed of many different cultures, each of which has different values and practices, and not all of which are entirely comprehensible or acceptable to each other, conflicts are inevitable. In the absence of a practice of intercultural dialogue, conflicts are insoluble except by the imposition of one culture's views on another. A society of cultural enclaves and de facto separatism is one in which different cultures do not know how to talk to each other, are not interested in each other's well-being, and assume that they have nothing to learn and nothing to gain from interaction. This becomes a problem for urban governance and for planning in cities where contact between different cultures is increasingly part of everyday urban life in the growing number of multiethnic neighbourhoods, in spite of the efforts of some groups to avoid 'cultural contamination' or ethnic mixture by fleeing to gated communities or so-called ethnic enclaves. A pragmatic argument, then, is that intercultural contact and interaction is a necessary condition for being able to address the inevitable conflicts that will arise in multicultural societies. Another way of looking at the question of why intercultural encounters might be a good thing would start with the acknowledgement that different cultures represent different systems of meaning and versions of the good life. But each culture realizes only a limited range of human capacities and emotions and grasps only a part of the totality of human existence: it therefore 'needs others to understand itself better, expand its intellectual and moral horizon, stretch its imagination and guard it against the obvious temptation to absolutize itself' (Parekh 2000: 336–7). These are arguments that will be further developed in what follows.

James Donald: an ethical indifference

In *Imagining the Modern City* (1999), Donald seems to take a less moralistic, less prescriptive, more pragmatic approach to the question of how we might live together. He is critical of the two most popular contemporary urban imaginings: the traditionalism of the New Urbanism (with its ideal of community firmly rooted in the past), and the cosmopolitanism of Richard Rogers, adviser to Tony Blair and author of a policy document advocating an urban renaissance, a revitalized and re-enchanted city (Urban Task Force 1999). What is missing from Rogers' vision, according to Donald, is 'any real sense of the city not only as a space of community or pleasurable encounters or self-creation, but also as the site of aggression, violence, and paranoia' (Donald 1999: 135). Is it possible, he asks, to imagine change that acknowledges difference without falling into phobic utopianism, communitarian nostalgia or the disavowal of urban paranoia?

Echoing Iris Young (1990), Donald sets up a normative ideal of city life that acknowledges not only the necessary desire for the security of home, but also the inevitability of migration, change and conflict, and thus an 'ethical need for an openness to unassimilated otherness' (Donald 1999: 145). He argues that it is not possible to domesticate all traces of alterity and difference. 'The problem with community is that usually its advocates are referring to some phantom of the past, projected onto some future utopia at the cost of disavowing the unhomely reality of living in the present' (Donald 1999: 145). If we start from the reality of living in the present with strangers, then we might ask what kind of commonality might exist or be brought into being. Donald's answer is 'broad social participation in the never completed process of making meanings and creating values . . . an always emerging, negotiated common culture' (Donald 1999: 151).

This process requires time and forbearance, not instant fixes. This is community redefined neither as identity nor as place but as a productive process of social interaction. Donald argues that we do not need to share cultural traditions with our neighbours in order to live alongside them, but we do need to be able to talk to them, while also accepting that they are and may remain strangers (as will we).

If this is the pragmatic urbanity that can make the violence of living together manageable, then urban politics would mean strangers working out how to live together. This is an appropriately *political* answer to Sennett's question of how multicultural societies might arrive at some workable notion of a common destiny, and foreshadows my later discussion of the importance of an agonistic democratic politics (Part 2). But when it comes to a thicker, more sociological description of this 'openness to unassimilable difference', the mundane, pragmatic skills of living in the city, sharing urban turf, neither Donald nor Sennett has much to say. Donald suggests:

> reading the signs in the street; adapting to different ways of life right on your doorstep; learning tolerance and responsibility – or at least, as Simmel taught us, indifference – towards others and otherness; showing respect, or self-preservation, in not intruding on other people's space; picking up new rules when you migrate to a foreign city.
>
> (Donald 1999: 167)

Donald seems to be contradicting himself here in retreating to a position of co-presence and indifference, having earlier advocated something more like an agonistic politics of broad social participation in the *never completed process* of making meanings and an *always emerging, negotiated common culture*. Surely this participation and negotiation in the interests of peaceful coexistence require something like daily habits of (perhaps quite banal) intercultural interaction in order to establish a basis for dialogue, which is difficult, if not impossible, without some pre-existing contact that can develop into trust. I now turn to Ash Amin for a discussion of how and where this daily interaction and negotiation of ethnic and other differences might be encouraged.

Ash Amin: a politics of local liveability

Amin's *Ethnicity and the Multicultural City. Living with Diversity* (2002) is a self-described 'think piece' that uses the 2001 race riots in three northern British cities (Bradford, Burnley and Oldham) as a springboard 'to discuss what it takes to combat racism, live with difference and encourage mixture in a multicultural and multiethnic society' (Amin 2002: 2). The dominant ethnic groups present in Bradford, Burnley and Oldham are Pakistani and Bangladeshi, of both recent and longer-term migrations. What this reflects is the twin and interdependent forces of postcolonialism and globalization, and these are Amin's starting points. As several scholars have pointed out (Sassen 1996; Rocco 2000), the contemporary phenomena of immigration and ethnicity are constitutive of globalization and are reconfiguring the spaces of and social relations in cities in new ways. Cultures from all over the world are being de- and re-territorialized in global cities, whose neighbourhoods accordingly become 'globalized localities' (Albrow 1997: 51). The spaces created by the complex and multidimensional processes of globalization have become strategic sites for the formation of transnational identities and communities, as well as for new hybrid identities and complicated experiences and redefinitions of notions of 'home'.

This is the context for Amin's (2002) interpretative essay on the civil disturbances, which he sees as having both material and symbolic dimensions. He draws on ethnographic research to deepen understanding of both dimensions, as well as to assist in his argument for a focus on the everyday urban, 'the daily negotiation of ethnic difference'. Ethnographic research in the UK on areas of significant racial antagonism has identified two types of neighbourhoods. The first are old white working-class areas in which successive waves of non-white immigration have been accompanied by continuing socio-economic deprivation and cultural and/or physical isolation 'between white residents lamenting the loss of a golden ethnically undisturbed past, and non-whites claiming a right of place' (Amin 2002: 5). The second are 'white flight' suburbs and estates that have become the refuge of an upwardly mobile working class and a fearful middle class disturbed by what they see as the replacement of a 'homely white nation' by foreign cultural contamination. Here, white supremacist values are activated to terrorize the few immigrants who try to settle there. The riots of 2001 displayed the processes at work in the first type of neighbourhood, but also the white fear and antagonism typical of the second type (Amin 2002: 2).

What is important to understand is that the cultural dynamics in these two types of neighbourhood are very different from those in other ethnically mixed cities and neighbourhoods where greater social and physical mobility, a local history of compromises, and a supportive local institutional infrastructure have come to support co-habitation (see Albrow 1997). In the northern mill towns that are the subject of Amin's reflection, when the mills declined, white and non-white workers alike were unemployed. The largest employers soon became the public services, but discrimination kept most of these jobs for whites. Non-whites pooled resources and opened shops, takeaways and minicab businesses. There was intense competition for low-paid and precarious work. Economic uncertainty and related social deprivation has been a constant for over 20 years and 'a pathology of social rejection . . . reinforces family and communalist bonds' (Amin 2002: 4). Ethnic resentment has bred on this socio-economic deprivation and sense of desperation. It is in such areas that social cohesion and cultural interchange have failed.

What conclusions does Amin draw from this? For one thing, he argues against several currently popular policy fixes. One such fix is based on the belief that cultural and physical isolation lies at the heart of the disturbances, so the way forward must lie in greater ethnic mixing in housing at the neighbourhood scale (see Home Office 2001). Another popular policy fix in the urban literature looks to the powers of visibility and encounter between strangers in the open or public spaces of the city. The freedom to associate and mingle in cafés, parks, streets, shopping malls and squares (a feature of Richard Rogers' recipe for urban renaissance) has been linked to the development of an urban civic culture based on the freedom and pleasure of lingering, the serendipity of the chance encounter, and the public awareness that these are shared spaces. The depressing reality, Amin counters, is that far from being spaces where diversity is being negotiated, these spaces tend either to be territorialized by particular groups (whites, youth, skateboarders, Asian families) or they are spaces of transit, with very little contact between strangers. 'The city's public spaces are not natural servants of multicultural engagement' (Amin 2002: 11).

If ethnic mixture through housing cannot be engineered, and public space is not the site of meaningful multicultural encounter, how can fear and intolerance be challenged, how might residents begin to negotiate and come to terms with difference in the city? Amin argues that the contact spaces of housing estates and public places fall short of nurturing interethnic understanding, 'because they are not spaces of interdependence and habitual engagement' (Amin 2002: 12). He goes on to suggest that the sites for coming to terms with ethnic (and surely other) differences are the 'micro-publics' where dialogue and prosaic negotiations are compulsory, in sites such as the workplace, schools, colleges, youth centres, sports clubs, community centres, neighbourhood houses and colleges of further education, in which people from different cultural backgrounds are thrown together in new settings which disrupt familiar patterns and create the possibility of initiating new attachments. Other sites include community gardens, child-care facilities, Neighbourhood Watch schemes, youth projects and the regeneration of derelict spaces. I provide just such an example (Sandercock 2003) in the Community Fire Station in the Handsworth neighbourhood of Birmingham, where white Britons are working alongside Asian and Afro-Caribbean Britons in a variety of projects for neighbourhood regeneration and improvement (the Collingwood Neighborhood House in Vancouver is another example of a successful site of intercultural interaction – see Dang 2002; Sandercock 2003). Part of what happens through such everyday contact is the overcoming of feelings of strangeness in the simple process of sharing tasks and comparing ways of doing things. But such initiatives will not automatically become sites of social inclusion. They also need organizational and discursive strategies that are designed to build a voice, to foster a sense of common benefit, to develop confidence among disempowered groups, and to arbitrate when disputes arise (Sandercock 2004b). The essential point is that 'changes in attitude and behaviour spring from lived experiences' (Amin 2002: 15).

The key policy implication from Amin's work, then, is that the project of living with diversity needs to be worked at 'in the city's micro-publics of banal multicultures' (Amin 2002: 13). Amin suggests a new vocabulary of local accommodation to difference – 'a vocabulary of rights of presence, bridging difference, getting along' (2002: 17). The achievement of these rights depends on a politics of active local citizenship, an agonistic politics (as sketched by Donald 1999 and Mouffe 2000) of broad social participation in the never completed process of making meanings, and an always emerging, contested and negotiated common culture.

The foregoing analysis of three reflections on multicultural urban coexistence offers a richer understanding of what a *social project of cosmopolitan urbanism* entails. It suggests a research and policy focus at the level of the neighbourhood, looking for and encouraging intercultural encounters and exchanges, inventing local institutions and designing public places that create the spaces for such interaction in the daily negotiations of difference that characterize urban life. Further, it is clear that merely creating spaces where intercultural exchange is encouraged is not enough to guarantee social inclusion. Organizational and discursive strategies are also necessary in order to build voice, to foster a sense of solidarity across differences, to develop confidence among disempowered groups, and to mediate when disputes arise. A recognition that conflict is inevitable and ineradicable is a good place to begin thinking about a twenty-first-century cosmopolitan urbanism, which leads to the following discussion in Part 2 of cosmopolitan urbanism as a *political* (as well as sociological) project.

Part 2: Cosmopolitan urbanism as a political project of intercultural coexistence

In proposing cosmopolitan urbanism as a political project for the twenty-first century, my starting point is an acknowledgement of at least three fatal flaws of twentieth-century multiculturalism: as a state-based project (Mitchell 1996; Scott 1998; Sandercock 2004a); as an ethno- and racially based approach grounded in a static understanding of culture (Bisoondath 2002; Mahtani 2002); and as a product of racialized Western liberal democracies living in an as yet unresolved postcolonial condition that confounds the best of liberal intentions (Bannerji 1995, 2000; Hage 1998; Henry *et al.* 2000; Hill 2001). Space prevents me from repeating these critiques here, but my conclusion is not that we should abandon the multicultural project. Rather, it needs to be rethought. I agree with Stuart Hall (2000) that 'the multicultural question' is both a global and local terrain of political contestation with crucial implications for the West. We are inevitably implicated in the politics of multiculturalism: that is, the actual production of multiculturalism on the ground – which I think of as the spatiality and sociality of immigration – given that we live in an age of globalization and global migrations. Therefore, we need to find a way to publicly manifest the significance of cultural diversity, and to *debate the value* of various identities/differences: that is, to ask (as Chantal Mouffe does) which differences exist, but should not, and which do not exist, but should (Mouffe 2000). The concept of multiculturalism needs to be transformed in response to critiques of its fatal flaws, rather than abandoned. This leads me to define an *intercultural perspective* (or a cosmopolitan urbanism) as a political and philosophical basis for thinking about how to deal with the challenge of difference in the mongrel cities of the twenty-first century.

Let me suggest five necessary and interrelated components of an intercultural perspective: the dialectics of identity/difference; the centrality of conflict, or an agonistic democratic politics; the right to difference; the right to the city; and a shared commitment to political community (my thinking here has been inspired by Connolly 1991; Tully 1995; Mouffe 2000; and Parekh 2000).

The paradoxical dialectics of identity/difference

We all grow up in a culturally structured world, are deeply shaped by it, and necessarily view the world from within a specific culture. The cultural embeddedness of humans would seem to be inescapable, and some form of cultural identity and belonging seems unavoidable. And yet we are capable of critically evaluating our own culture's beliefs and practices, and of understanding and appreciating, as well as criticizing, those of other cultures. We are capable of imagining and desiring cultural change. No culture is perfect or can be perfected, but all cultures have something to learn from and to contribute to others. Intercultural dialogue is thus a necessary component of cultural growth and development.

To some extent, one's own cultural identity is and will always be defined in relation to degrees of difference from others. And yet no culture is entirely static. Cultures are always evolving, dynamic and ultimately hybrid, containing multiple differences within themselves that destabilize rigid understandings of identity. And since diversity exists within as well as between cultures, no pure form of cultural identity is capable of being the foundation of membership in a political community. This implies the 'end of mainstream', in politico-cultural terms, and the birth of plurality, as the basis of political community.

An agonistic democratic politics

In demographically multicultural societies and polities, conflicts over values and lifestyles, ways of being and ways of knowing, are unavoidable. As long as there is global movement of peoples and their accompanying cultural baggage, consensus will only ever be temporary, as each newcomer/group engages in the political arena in an attempt to redefine the society in its own image. An agonistic politics entails broad social participation in the never completed process of making meanings and creating values. An agonistic politics implies 'the end of mainstream' in terms of the end of a single dominant culture in any polity, perpetual contestation over what is or might become common ground, and negotiation towards a sense of shared destiny. The quest for such common ground and shared destiny should not ordinarily subsume the right to difference, but that right is also a matter of political negotiation, a component of an agonistic politics.

The right to difference

As a daily political practice interculturalism recognizes the right to difference, expressed as the legitimacy and specific needs of minority and subaltern cultures. However, the right to difference at the heart of cosmopolitan urbanism must be perpetually contested against other rights – human rights, for example – and redefined according to emerging considerations and values. The right to difference must always be tempered by the imperative of peaceful coexistence and the recognition of shared societal and global challenges such as ecological sustainability and social justice.

The right to the city

In a world that will be predominantly urban by the middle of the twenty-first century, negotiating peaceful intercultural coexistence, block by block, neighbourhood by neighbourhood, will become a central preoccupation of citizens as well as urban professionals and politicians. The right to the city is the right of all residents to presence throughout the city, the right to inhabit and appropriate public space, and the right to participate as an equal in public affairs, to be engaged in debating and designing the future of the city and creating new intercultural spaces and built forms, and new ways of being together in the city.

A shared commitment to political community

A sense of belonging in an intercultural society cannot be based on race, religion, ethnicity or any other such marker of identity/difference. Rather, that sense of belonging must be based on a shared commitment to political community, and specifically to a political community founded on the principles of an agonistic democratic politics. Such a political community remains perpetually open to redefinition as its membership changes, but there must be agreed-on procedures for debate and for resolving conflicts, and there must be legal and institutional protections against discrimination. A shared commitment to a political community also requires an empowered citizenry, which in turn means addressing prevailing inequalities of political and economic power as well as developing new stories about and symbols of national and local identity and belonging.

There are (at least) two public goods embedded in an intercultural or cosmopolitan urbanism, based on these five components of a political philosophy. One is the critical freedom to question in thought, and challenge in practice, one's inherited cultural ways. The other is the recognition of the widely shared aspiration to belong to a culture and a place, and so to be at home in the world (Tully 1995). This sense of belonging would be lost if one's culture were excluded, or if it was imposed on everyone. But there can also be a sense of belonging that comes from being associated with other cultures, gaining in strength and compassion from accommodation among and interrelations with others, and it is important to recognize and nurture those spaces of accommodation and intermingling. This concept of interculturalism accepts the indispensability of group identity to human life (and therefore to politics), precisely because it is inseparable from belonging. But this acceptance needs to be complicated by an insistence, indeed a vigorous struggle against the idea that one's own group identity has a claim to intrinsic truth. If we can acknowledge a drive within ourselves, and within all of our particular cultures, to naturalize the identities given to us, we can simultaneously be vigilant about the danger implicit in this drive, which is the almost irresistible desire to impose one's identity, one's way of life, one's very definition of normality and of goodness, on others. Thus we arrive at a lived conception of identity/difference that recognizes itself as historically contingent and inherently relational; and a cultivation of a care for difference through strategies of critical detachment from the identities that constitute us (Connolly 1991; Tully 1995). In this intercultural imagination, the twin goods of belonging and of freedom can be made to support rather than oppose each other.

From an intercultural perspective, the good society does not commit itself to a particular vision of the good life and then ask how much diversity it can tolerate within the limits set by this vision. To do so would be to foreclose future societal development. Rather, an intercultural perspective advocates accepting the reality and desirability of cultural diversity and then structuring political life accordingly. At the very least, this political life must be dialogically and agonistically constituted. But the dialogue requires certain institutional preconditions, such as freedom of speech, participatory public spaces, empowered citizens, agreed procedures and basic ethical norms, and the active policing of discriminatory practices. It also calls for 'such essential political virtues as mutual respect and concern, tolerance, self-restraint, willingness to enter into unfamiliar worlds of thought, love of diversity, a mind open to new ideas and a heart open to others' needs, and the ability to live with unresolved differences' (Parekh 2000: 340).

Since commitment, or belonging, must be reciprocal, citizens will not feel these things unless their political community is also committed to them and makes them feel that they belong. Here's the rub, then. An intercultural political community 'cannot expect its members to develop a sense of belonging to it unless it equally values and cherishes them in all their diversity, and reflects this in its structure, policies, conduct of public affairs, self-understanding and self-definition' (Parekh 2000: 342). It would be safe to say that no existing (self-described) multicultural society can yet claim to have achieved this state of affairs. But in recent years these issues have been identified, increasingly documented, and are becoming the focus of political activity in many countries (see Sandercock 2003).

Empirically speaking, the twenty-first century is indisputably the century of multicultural cities and societies. This means it will also inevitably be the century of struggle for interculturalism, and against fundamentalism, which is a belief in cultural (or religious) purity. A cosmopolitan urbanism then, or cosmopolis – to use an earlier term of mine (Sandercock 1998) – is a utopian social and political project for negotiating the socio-cultural transformations of human settlements in the coming age. It is a love song to our mongrel cities, rather than a war against them.

References

Albrow, M. (1997) 'Travelling beyond local cultures: socioscapes in a global city', in J. Eade (ed.), *Living the Global City. Globalization as a Local Process*, London: Routledge.

Amin, A. (2002) *Ethnicity and the Multicultural City. Living with Diversity*, Report for the Department of Transport, Local Government and the Regions, Durham: University of Durham.

Bannerji, H. (1995) *Thinking Through*, Toronto: Women's Press.

Bannerji, H. (2000) *The Dark Side of the Nation: Essays on Multiculturalism, Nationalism and Gender*, Toronto: Canadian Scholars' Press Inc.

Beck, U. (1999) 'Democracy beyond the nation-state: a cosmopolitan manifesto', *Dissent* (Winter): 53–5.

Bisoondath, N. (2002) *Selling Illusions: The Cult of Multiculturalism in Canada*, Toronto: Penguin Books.

Brennan, T. (1997) *At Home in the World: Cosmopolitanism Now*, Cambridge, MA: Harvard University Press.

Calhoun, C. (2002) 'The class consciousness of frequent travellers: towards a critique of actually existing cosmopolitanism', in S. Vertovec and R. Cohen (eds), *Conceiving Cosmopolitanism. Theory, Context, and Practice*, Oxford: Oxford University Press.

Cheah, P. and Robbins, B. (eds) (1998) *Cosmopolitics: Thinking and Feeling Beyond the Nation*, Minneapolis, MN: University of Minnesota Press.

Cohen, J. and Nussbaum, M. (1996) *For Love of Country: Debating the Limits of Patriotism*, Boston, MA: Beacon Press.

Connolly, W. (1991) *Identity\Difference*, Ithaca, NY: Cornell University Press.

Dang, S. (2002) 'Creating Cosmopolis: The End of Mainstream', Master's thesis, School of Community and Regional Planning, University of British Columbia, Vancouver.

Donald, J. (1999) *Imagining the Modern City*, London: Athlone Press.

Falk, R. (2000) *Human Rights Horizons: The Pursuit of Justice in a Globalizing World*, New York: Routledge.

Hage, G. (1998) *White Nation: Fantasies of White Supremacy in a Multicultural Society*, Sydney: Pluto Press.

Hall, S. (2000) 'Conclusion: the multi-cultural question', in B. Hesse (ed.), *Un/settled Multiculturalisms*, London: Zed Books.

Harvey, D. (2000) 'Cosmopolitanism and the banality of geographical evils', *Public Culture*, 12: 529–64.

Henry, F., Tator, C., Mattis, W. and Rees, T. (2000) *The Color of Democracy: Racism in Canadian Society*, Toronto: Harcourt Canada.

Hill, L. (2001) *Black Berry, Sweet Juice: On Being Black and White in Canada*, Toronto: HarperCollins.

Hollinger, D. (2002) 'Not universalists, not pluralists: the new cosmopolitans find their own way', in S. Vertovec and R. Cohen (eds), *Conceiving Cosmopolitanism*, Oxford: Oxford University Press.

Home Office (2001) *Building Cohesive Communities: A Report of the Ministerial Group on Public Order and Community Cohesion*, London: Home Office/Her Majesty's Government.

Mahtani, M. (2002) 'Interrogating the hyphen-nation: Canadian multicultural policy and "mixed race" identities', *Social Identities*, 8: 67–90.

Mitchell, K. (1996) 'In whose interest? Transnational capital and the production of multiculturalism in Canada', in R. Wilson and W. Dissanayake (eds), *Global/Local. Cultural Production and the Transnational Imaginary*, Durham, NC and London: Duke University Press.

Mouffe, C. (2000) *The Democratic Paradox*, London: Verso.

Nussbaum, M. (1996) 'Patriotism and cosmopolitanism', in M. Nussbaum and J. Cohen (eds), *For Love of Country: Debating the Limits of Patriotism*, Boston, MA: Beacon Press.

Parekh, B. (2000) *Rethinking Multiculturalism*, London: Macmillan.

Rocco, R. (2000) 'Associational rights-claims, civil society and place', in E. Isin (ed.), *Democracy, Citizenship and the Global City*, London: Routledge.

Rushdie, S. (1992) *Imaginary Homelands*, London: Granta Books.

Sandercock, L. (1998) *Towards Cosmopolis. Planning for Multicultural Cities*, Chichester: Wiley.

Sandercock, L. (2003) *Cosmopolis 2: Mongrel Cities of the 21st Century*, London: Continuum Books.

Sandercock, L. (2004a) 'The angel of progress falls to earth (again): interrogating the Canadian multicultural project', keynote paper presented at the annual conference of the American Association of Geographers, Philadelphia, March.

Sandercock, L. (2004b) 'Constructing new ways of living together in multicultural cities: Vancouver's Collingwood Neighborhood House', paper presented to Ninth Annual Metropolis Conference, Geneva, September.

Sassen, S. (1996) 'Whose city is it? Globalization and the formation of new claims', *Public Culture*, 8: 205–23.

Scott, J. C. (1998) *Seeing Like a State. How Certain Schemes to Improve the Human Condition Have Failed*, New Haven, CT: Yale University Press.

Sennett, R. (1994) *Flesh and Stone. The Body and The City in Western Civilization*, New York: Norton.

Tully, J. (1995) *Strange Multiplicity. Constitutionalism in an Age of Diversity*, Cambridge: Cambridge University Press.

Urban Task Force (1999) *Towards an Urban Renaissance*, London: Spon.

Van der Veer, P. (2002) 'Colonial Cosmopolitanism', in S. Vertovec and R. Cohen (eds), *Conceiving Cosmopolitanism*, Oxford: Oxford University Press.

Vertovec, S. and Cohen, R. (eds) (2002) *Conceiving Cosmopolitanism*, Oxford: Oxford University Press.

Young, I. M. (1990) *Justice and the Politics of Difference*, Princeton, NJ: Princeton University Press.

Zachary, G. P. (2000) 'Duelling multiculturalisms: the urgent need to reconceive cosmopolitanism', Oxford: ESRC Transnational Communities Programme, Working Paper WPTC-2K-04.

The paradox of cosmopolitan urbanism: rationality, difference and the circuits of cultural capital

Gary Bridge

Introduction

Cosmopolitan urbanism is, I argue, the conjunction of particular forms of professional, rational knowledge and its acquisition in the spaces of key metropolitan centres. In its avowed openness to difference, cosmopolitan knowledge is allied with current ideas of transversal rationality as a logic of transition between spaces and cultures, rather than a traditional rationality based on hierarchy and fixed location. Rather than this cosmopolitan knowledge being based on 'decontextualised cultural capital' (Hannerz 1996: 108), however, I argue that it relies on the time-space particularities of the acquisition and reproduction of the various strands of cultural capital. Thus the circuits of cultural capital that make cosmopolitan knowledge possible are often antithetical to cosmopolitanism as a form of openness to difference. The drive to keep all social fields in play results in a narrowing of the social field and of the encounter with others. The more coherence there is between the circuits of economic and cultural capital, the more likely it is that the field is socially elite and exclusive, and this militates against the ideal of cosmopolitanism being open to otherness. I explore the paradoxes of cosmopolitan rationality using the examples of elite gentrification in Sydney and London, and 'provincial' gentrification in Bristol, England.

Cosmopolitanism, transversal rationality and smooth space

There is a growing literature on the associations between professionalism, rationality and the rise of a global new middle class (Gouldner 1979; Ehrenreich and Ehrenreich 1979; Hannerz 1996; Isin 2000; Robbins 2001). Alvin Gouldner (1979) argued that the defining characteristic of the new class was its cosmopolitanism. Hannerz (1990) develops the distinction between locals and cosmopolitans, with the influence of the former being based not on what they know but who they know, whereas for cosmopolitans the relationship is reversed.

> Cosmopolitans . . . base[d] whatever influence they had on a knowledge less tied to particular others, or to the unique community setting. They came equipped with special knowledge, and they could leave and take it with them without devaluing it . . . they are the 'new class', people with credentials, decontextualised cultural capital.
>
> (Hannerz 1996: 108)

Their knowledge is accompanied by a set of analytics – 'a culture of critical discourse' (Gouldner 1979). Hannerz (1990, 1996) argues that this critical discourse is a form of metacommunication that is reflexive, problematising and explicit, in contrast to commonsense knowledge that is tacit, ambiguous and contradictory.

A further element in the discussion of professionalism, rationality and the new middle class is the assumption that cosmopolites demonstrate an openness to difference, and that the deployment of their rationality is marked by its sensitivity to difference. Recent writing on cosmopolitanism (Hannerz 1996; Dharwadker 2001; themed issue 19 of *Theory, Culture and Society* 2002) seems to stress that it is ideally defined by mobility and a sense of openness to otherness:

> Of importance for cosmopolites is openness to the other, as Gilles Deleuze repeatedly puts it, to 'think towards the horizon'; to alter one's perceptions and to invent new 'smooth spaces' instead of building walls, barriers and prisons to conserve one's interests.
>
> (Conley 2002: 131)

Conley (2002) ties this to the type of rhizomatic connections that Deleuze and Guattari postulate. To be a cosmopolite is to occupy a junction in a rhizomatic network where there are only lines of flight to other diverse connections. These connections are themselves only temporary and partial in a way that Deleuze defined as chaosmopolis (cited in Conley 2002) in which a transversal rationality operates.

Transversal rationality is a contemporary manifestation of cosmopolitan knowledge involving expertise and openness to difference. Adapting Welsch's (1998) arguments, the history of modernity is the emptying out of the idea of reason. First to fall was a belief in the unitary content of reason in the form of universal truths. In late modernity there were attempts to rescue reason as procedure, or the ways of getting at the truth. This was done most famously by Habermas (1984, 1987) in his argument that the inherent telos of language contained claims to validity. The procedures for the defence of validity claims were universal (the good reason or grounds for argument – communicative rationality) even if the content differed. But with the postmodern turn even calls to the universality of procedure were found to be culturally specific. For many this meant the abandonment of reason. Welsch (1998) defends one final element of reason which he calls 'pure faculty'. Universal reason has separated out into different spheres of rationality (over action, aesthetics and morality, for example). The realisation that there are non-compatible spheres of rationality is itself a critical faculty: the capacity for reason. This is a reason that moves between rationalities and does not try to unite them.

Schrag's (1992) idea of transversal rationality also resonates with this sense of cosmopolitanism. It arises out of community context but involves the capacity to be distant from community norms. Transversality avoids the verticality or hierarchy of traditional rationality as the logic of identity, to be replaced by the horizontality of a logic of transition. It cuts across difference obliquely. Schrag (1992) calls on Bakhtin's idea of chronotope as the socio-practical constitution of time and space through dialogue. Rationality is 'the ability to discern and articulate how our discursive and non-discursive practices hang together in the texture of everyday life' (Schrag 1992: 94). In the relation with 'the Other', that key moment of cosmopolitanism, there is a claim to 'convergence without coincidence' (Schrag 1992: 158).

Dominant contemporary manifestations of cosmopolitanism involve knowledge based on professional competences and transnational connections. The degree to which this is a movement towards a global democracy, or a new transnational exclusivity of elites, is at the core of these debates, as are questions about whether the openness to the Other is genuine or a kind of fake, consumption-oriented transnational flâneurism (May 1996; Hage 1997). Also central are questions about the degree to which cosmopolitanism is or is not tied up with cities. It is a form of knowledge and conduct that, in its concentrations and connections, might be associated with certain urban spaces, especially particular districts of global cities (Sassen 2000).

Elite enclaves in global cities are very much part of the infrastructure of what Castells (1996) calls 'the space of flows'. The space of flows is 'the material organisation of time-sharing social practices that work through flows' (Castells 1996: 412), where flows are defined as 'purposeful, repetitive, programmable sequences of exchange and interaction between physically disjointed positions held by social actors in the economic, political, and symbolic structures of society'. The space of flows is supported by circuits of electronic impulses, involving microelectronics, telecommunications and so on and by nodes and hubs in these networks, of which the 'global city' is a key instance, and by the spatial organisation of the transnational managerial elite.

The spatial organisation of the dominant elites form landscapes of power (Zukin 1991) that provide a seamless environment for work and leisure all over the world. It involves a symbolic economy that excludes the masses by the enormous amounts of economic and cultural capital required for membership. The symbolic economy is a manifestation of 'good taste' in all aspects of life, from body presentation and deportment, through exercise, diet, desirable residential address, home furnishings, in a set of aesthetic dispositions that are increasingly transnational. These elements of lifestyle symbolic capital are supported when cosmopolites are on the move, via airport VIP lounges, exclusive hotels with familiar ambiances and technical and personnel support to facilitate constant communication and a comfortable working environment.

This capacity to move between contexts without seeking to encompass difference is what we might think of as a cosmopolitan characteristic. This is especially so when we consider Welsch's argument that the world is becoming increasingly transcultural (Welsch 1995). Rejecting multiculturalism for its still-essentialised notion of culture, transculturalism proposes that cultural boundaries and subjectivities are breaking down and becoming hybrid. So we have transversal reason as a cosmopolitan capacity in an increasingly transcultural world. Transversal reason shifts between rationalities that themselves are breaking down into a hybridised world: the logic of transitions in smooth space.

Cosmopolitanism qualified: the rough space of cultural capital

Cultural capital takes three forms: institutionalised cultural capital or formally accredited learning; objectified cultural capital, such as art, books and the stylistic aspects of interior decoration and furniture; and embodied cultural capital, or the non-accredited and sometimes tacit knowledge, tastes and dispositions absorbed through participation in a particular habitus.

The smooth space of cosmopolitanism can be seen as that of the decontextualised cultural capital of the cosmopolite. Hannerz (1990, 1996), following Gouldner (1979) and others, equates cultural capital with the professional critical knowledge that the cosmopolite embodies and is able to transport anywhere. Yet various strands of cultural capital as currently embodied in the new middle-class professional worker have a history of acquisition and institutional settings. These strands also have relationships with economic and social capital and there are key questions here about translation between capitals and their possible contradictions. Smooth space can be seen as a form of coherence in the relations between material, social capital and the various strains of cultural capital. I suggest a number of contradictions and discontinuities between circuits of capital and how they relate to an idea of cosmopolitanism. I argue this through the example of gentrification.

At this point I think we need to make explicit the connections between the content of professional expertise and the mode of its acquisition and deployment. What makes the professional knowledge of the new middle class translatable? One part is its technical aspect – it specialises in systematising the objects of knowledge. Another equally important part, however, is its ability to legitimise itself as the dominant deployment of knowledge. This is the element of cultural capital – the deployment of the resources of the habitus in constructing the metric of judgement over objects and situations – being able to discern between the categories of that metric and then to legitimise this system of classification as the legitimate system – a form of social power.

The legitimisation of the professional form of knowledge is consolidated by the same principles of judgement being extended to the whole sphere of personal conduct, lifestyle and taste, a habitus of cosmopolitan knowledge. It might at first seem a little odd to be using an idea of habitus in relation to cosmopolitanism. Habitus has been critiqued as a notion very much tied to national social structures and as a rather static notion of social space. Yet Hillier and Rooksby (2002) found the continued relevance of the idea of habitus in a globalised world. Habitus need not be tied to a national context, particularly in its relation to Bourdieu's idea of a field as a relational and discursive force field of social space (Isin 2000).

Cultural capital, cosmopolitanism and gentrification

There is much in the gentrification literature to suggest strong connections with the ideas of cosmopolitanism and habitus, especially through debates about global cities and globalisation (Atkinson and Bridge 2005). There is considerable overlap between the occupational categories of professional knowledge workers that are involved in gentrification and those highly mobile knowledge professionals that have diverse transnational connections identified in debates on cosmopolitanism. The desire to live in the central city would seem to signal a disposition towards social and cultural diversity in which the city itself becomes emblematic of a wider set of cosmopolitan activities. Indeed Podmore (1998) has argued that loft living connotes a set of consumption and lifestyle orientations that meet ready reception transregionally and transnationally because of a shared habitus. In the case of the gentrification of loft spaces, Podmore (1998: 287) adapts Hannerz's (1990) arguments:

> Loft dwellers, whether they are artists or corporate executives, generally have high levels of cultural capital; they are cultural elites, physically located in specific urban environments but more broadly connected to a global habitus of shared dispositions and social practices through the mass media and other communications technologies.

Similar observations have been made by Ley (1996) and Mills (1988) about a postmodern liberal politics of liveability that has transformed the central cities of many Western nations as a result of gentrification activity. The ubiquity of waterfront executive apartments, lofts and even the brownstones (Eastern USA), whitepainted (Canada), Victorian villas (London) and Victoriana (Melbourne) show a good deal of consistency across countries in owner rehabilitation of older properties, or sweat equity gentrification.

On the other hand, there is some evidence that the gentrification habitus is superficial or strongly marketing led. Thus, as Mills (1988) suggests in 'Life on the upslope', and as Zukin (1991) observes, there is a critical infrastructure of gentrification that involves a discursive community including restaurant critics and lifestyle gurus that support the gentrification neighbourhood activities. Alternatively its superficiality is registered by suggestions that its openness to otherness is more a kind of aesthetic of consumption rather than any genuine sense of encounter and alterity. This parallels the discussions of superficial cosmopolitanism and is captured by May's (1996) and Hage's (1997) observations about what we might call the thin multiculturalism of gentrifiers which is limited to diverse cuisine and neighbourhood aesthetics.

Some of these debates might be helped by considering cultural capital not as decontextualised but in terms of the time-space patterns of its acquisition and deployment. Differences have been indicated by arguments suggesting the fragmentation of the new middle class rather than its coherence. This was presaged in Gouldner's (1979) original distinction between 'humanistic intellectuals' and the 'technical intelligentsia' of the new class. These fractions can be distinguished by the relative rates and styles of deployment of cultural and material capital. In what remains I want to suggest that some of the processes that start to fragment the new middle class and the gentrification aesthetic have resonance for considerations of what it means to be cosmopolitan.

An understanding of the internal dynamics of cultural capital at the neighbourhood level and its interrelations with social and economic capital is important here. There are situations in which there is a consolidation of cultural capital in the urban landscape, but this reduces the 'openness' of cosmopolitanism. I draw on examples from my own work on elite gentrification in Sydney, Australia (Bridge 2001a) as well as Butler (1997) and Butler and Robson's (2001, 2004) work on London to illustrate this point. Equally there are instances in certain neighbourhoods in non-global cities where the various elements of cultural capital might be in conflict with each other, leading to an attenuation of cultural capital again marking the limits of cosmopolitanism (Bridge 2003).

Cultural capital and the limits of cosmopolitanism: gentrification in Sydney, London and Bristol

Cultural capital does not exist in a social vacuum or smooth space. There are forces that ground its circulation and tie it very much to local passions and loyalties. This rougher space-time of cultural capital is suggested in a number of ways. One is in the relationship between cultural capital and other forms of capital. Thus in my study of gentrification and cultural capital in Sydney (Bridge 2001a) it was evident that, at the upper end of the gentrification market (where cosmopolites are most likely to be represented), there was an increasing performative contradiction between the demands of cultural capital through aesthetic display in housing and the logic of economic capital in investment strategies. Gentrification elites in Sydney were overinvesting economic capital to sustain cultural distinction in an increasingly competitive taste market. This was evident in the price paid for deceased estates as opposed to more recent conversions that were structurally more suitable but cosmetically off-putting. It also showed up in the ever more elaborate internal restructurings (often architect-designed) of the properties, especially to capture prestigious harbour views. This competitive aesthetic flowed out from the gentrifier-occupied property to include their purchase and rehabilitation of investment properties, which in terms of the rental market of Sydney were overcapitalised to maintain the elite gentrification aesthetic in the rental property.

As well as the competitive momentum to maintain distinction there are various taste boundaries that come with the aesthetic of gentrification that are antithetical to any wider 'cosmopolitan' acknowledgement of difference. One manifestation of this is the association of the gentrification aesthetic with whiteness: in the case of Sydney, an Anglo-Australian aesthetic of Victoriana (Jager 1986; Bridge 2001a). Any infringement of these taste limits was severely penalised in the housing market. This shows how cultural capital relates to economic capital and operates as a powerful force to exclude, or marginalise different cultural practices and tastes. This is described in stark terms by one Sydney estate agent:

> We sold a house in 'Cranston' Avenue once which was absolutely a Mediterranean home. When I say Mediterranean it was owned by an Italian chap who spent an absolute fortune, maybe A$300,000 on this beautiful home, and he'd ruined it. He'd taken the timber floors out, he'd fully tiled the property completely throughout, he'd taken all the timber windows and put aluminium windows in, he'd put fountains and made it look just out of character with Glebe – if it was Balkan Hills or if it was in Fairfield he would have sold it thirty times a day but because of the type of people who are buying – they said it's unrenovated – we've got to restore it – we've got to get it back to the Victorian style.

> GB: They saw it as an unrenovated property?
>
> An unrenovated home. And my problem with him was talking to him and saying . . . unfortunately the people who buy your house have to rip it out and start again. And he found it very hard to come to grips with that because he'd spent a fortune in time doing it up to his own taste, which wasn't the taste of the local people.

However transversally rational the knowledge of the new middle class may be, the performative legitimation of that rationality involves the deployment of cultural capital that in this case excludes difference.

The time-space co-ordinates in the acquisition, deployment and reproduction of cultural capital again suggest that it is not decontextualised, even for cosmopolites. Education is a key element of incorporated cultural capital, in terms of both its institutional aspects and its part in a wider repertoire of dispositions that are incorporated onto the educated body. The reproduction of cultural capital intergenerationally via the schooling strategies of cosmopolites is therefore of especial interest. There are those cosmopolites of a traditional type whose internationalism extends to the schooling of their children (the children of diplomats or senior business executives educated in international schools), and so in some senses their reproduction of cultural capital is decontextualised. But for others the intergenerational reproduction of cultural capital is a much more complex set of processes. This is clear from Butler (1997) and Butler and Robson's (2001, 2004) work on the geography of gentrification and circuits of schooling in London. Butler and Robson's work drew on Ball, Bowe and Gerwitz's (1995) research that linked 'cosmopolitan' parents and the careful construction of their child's school career. What Robson and Butler's work suggests is that cosmopolites and metropolitan locals alike are implicated in a highly localised set of strategies around certain neighbourhood schools to get their children off to an acceptable start on their educational careers. Housing moves to get into the right catchments, appeals over school allocation and catchment boundaries, and the appropriation of certain promising primary schools, through a high level of parent investment, are all in play. White middle-class residents of Brixton, London, whilst claiming to value the multicultural, cosmopolitan nature of the district, showed far more anxiety about the prospects for schooling and their time horizons for staying in the neighbourhood. In Telegraph Hill, south London, by contrast, the middle-class residents had grafted out a possible circuit of schooling which resulted in longer time horizons but also a narrower socio-demographic environment. The structure of services (amongst other things) and the nature of the city become important when we consider circuits of capital.

Butler and Robson (2001, 2004) situate their work in a broader set of arguments about the degree to which gentrification is a coping strategy by the new middle class to deal with the exigencies of globalisation – namely a more flexible and unpredictable labour market, long working hours and both partners working. Even these relatively privileged high-end service class workers are feeling the pressure of these workplace demands on their personal lives, particularly with respect to bringing up children. They fear that although their children may be well provided for materially, they lack the sort of security and emotional support they associate with their own childhoods. What Butler and Robson are suggesting, I think, is that this anxiety is resulting in all kinds of what could be called strategies of localisation. These consist of investment or interest in neighbourhood and local community, local schooling and a certain nostalgia evidenced through the gentrification aesthetic. Butler and Robson suggest that this explains the variegated geography of gentrification in London. An extended theoretical account of the conflicted subjectivities and vulnerabilities of cosmopolitan groups is given in a recent paper by David Ley (2004).

In terms of cultural capital I think these strategies of localisation can result in tensions that start to stretch and fray the idea of some kind of consistent gentrification habitus as a secure reproduction of cultural capital for some of the new urban middle class. This is especially the case where two of the elements of cultural capital – incorporated and objectified – start to pull against each other.

In my recent research in Bristol, UK, I have investigated the relationship between housing and neighbourhood aesthetics, cultural capital and the lifecycle. Bristol is a provincial city in south-west England but one with a high proportion of financial services, media and media-related and hi-tech industries. It has a similar occupational structure to London. By interviewing professionals with properties on the market in a gentrified inner urban neighbourhood, and then re-interviewing those that succeeded in moving, the study sought to capture the degree to which the cultural strategies that the gentrification aesthetic is assumed to represent are carried through to the next move (Bridge 2003). The study makes no claims about cosmopolites as such (although a number of the interviewees fell into the liberal professional/aesthetic sector with transnational connections). What it does reveal, however, is a diversity of trajectories through the gentrified neighbourhood. Middle-class movers within the gentrified neighbourhood encompassed a range of gentrifier types. There was the corporate gentrifiers who worked in the professional private sector, had lived in the neighbourhood a short time and were moving into the ready renovated gentrified properties. There were also community-oriented gentrifiers with longer-standing links in the neighbourhood, who had been more involved in the gentrification of their homes (including sweat equity gentrification). There were also marginal gentrifiers (Rose 1984) who valued the convenience of an inner urban location but who were stretched in terms of their housing costs and ability to renovate. Gentrifiers also crossed the range of liberal/intelligentsia and management/technical workers.

Of those leaving the neighbourhood for contrasting housing aesthetics elsewhere all were reluctant quitters. This is largely because the moves came out of practical constraints rather than aesthetic choice. The main constraint is the lack of accommodation 'suitable' for the logistics of child-rearing or the 'right' location for schooling (Bristol state schools are on the whole of low standard). In London the different socio-demographic trajectories of gentrifiers (especially the division between public sector and private sector professionals) sorts itself out into distinct neighbourhood types and circuits of schooling. In Bristol, however, gentrification is much more constrained, involving different trajectories within the same neighbourhood, and an exit from the city for child-rearing, suggesting a more provincial and tentative form of gentrification. This constrained gentrification has much to say about the relationship of space to fields of cultural capital and habitus. The reluctant movers show the difficulty of keeping all social fields (in a Bourdieuan sense (Bridge 2001b)) in play at the same time. They are forced to trade the current deployment of cultural capital in aesthetic display (objectified cultural capital) for a longer-term investment in the reproduction of cultural capital through the schooling of their children (incorporated cultural capital). The particular structural and spatial arrangement of these fields in different cities is thus crucial to understanding the contextual nature of cultural capital. In the Bristol case it is becoming more diffuse in as much as aesthetic display and intergenerational reproduction cannot be assured at the same time. The perceived necessity to secure good schooling pulls some of the gentrifiers back into a much more conventional middle-class housing career involving a move to the suburbs or satellite villages and into new build or other housing that does not have the gentrification cachet.

Middle-class strategies over education in the English context also reveal something more of the interrelationship between the idea of mobile or transversal rational knowledge and cultural capital in relation to ideas of cosmopolitanism. In England there is a quasi-market in education with a policy emphasis on parental choice and the comparison of schools as products involving the publication of league tables of pupil performance in certain key tests. Butler and Robson (2001) note how the middle classes tend to envision their children's education in terms of education careers. There is evidence to suggest how middle-class use of the league table results feeds in strongly to the housing market via desirable school catchments and other strategies involving pupil commutes that aim to secure a good school. In this sense that rationality of middle-class knowledge operates in the comparison of league tables and the abstraction of the judgement of children's education away from the local context but that then results in a number of highly localised strategies (e.g. school capture, house move, bussing children, private education) to cope with the consequences of the application of rational knowledge. Thus there is an increasing reflexivity in the use of abstract knowledge over schools that relates directly to the way that cultural capital reproduction becomes associated with certain neighbourhoods, or where acting on the knowledge can lead to all kinds of localised strategies to secure cultural capital reproduction amongst 'people like us' (as Butler and Robson (2001) describe it). The operation of abstract or decontextualised rational knowledge leads to much rougher spaces of cultural capital, fully contextualised and often socially divisive.

If Bristol represents the lower, provincial end of these processes, they are by no means absent in larger, even global cities. The time-space patterns of the acquisition, deployment and reproduction of cultural capital (in the examples given in this chapter) reveal cultural activity that is at times intensely localised. That localisation, however, does not necessarily result in a local engagement with unlike others. Butler and Robson (2001) characterise the sorts of relations that exist between gentrifiers and others, especially in the 'cosmopolitan' neighbourhood of Brixton, as resembling 'social tectonics'. I suggest it is helpful to think in terms of time-space rhythms. The objectified element of cultural capital here relates to space, through the aesthetics of gentrification as the habitus of a distinct class fraction (a new urban middle class) expressed through housing, the retail landscape and neighbourhood milieu. In contrast the incorporated element of cultural capital relies on longer-term temporal rhythms of investment via education at primary, secondary and tertiary level. At critical points the temporal undertow of incorporated cultural capital can pull down on the objectified and spatial aesthetic display of gentrification.

It is in the context of a global city like London that the tensions within cultural capital might be successfully resolved in terms of both aesthetics and education as a result of the scale of the metropolis and the range of neighbourhoods available, giving London its own geography of gentrification. It is only in this kind of location that one can envisage the successful reproduction of a distinct fraction of the middle class that we might call the new urban middle class. This is a cycle of class reproduction from the purchase of the first gentrified property, through a sustained gentrification trajectory (perhaps involving several neighbourhoods), incorporating childrearing and the education careers of children up to graduation from an 'appropriate' university. In the case of Bristol that kind of reproduction over the gentrification/education cycle cannot be sustained in the same way as it is in London, or in the same numbers. Butler and Robson (2001) are right to identify what they call a 'metropolitan habitus' of gentrification. Within that metropolitan habitus I interpret their work to be suggesting a series of mini-habituses, with a range of neighbourhood and education trajectories that break down initially in terms of distinctions between highly paid private sector professional employees and lower paid public sector professional employees.

Yet the metropolitan habitus that Butler and Robson (2001) define is unlikely to be a cosmopolitan one. The drive to keep all social fields in play (crudely, aesthetics and education) results more and more in a narrowing of the social field and the encounter with others. The more coherence there is between the circuits of economic and cultural capital, the more likely it is that the field is socially elite and exclusive, and this works against the ideal of cosmopolitanism being open to otherness.

The cosmopolitanism of urban situations

It might be precisely in the inability to have all the spheres of capital run smoothly, or in realms where cultural practice is not legitimised in the form of cultural capital, that openness to the Other may have its greatest potential (and pose the greatest threat). Here I turn to a historical example. The work of the pragmatist John Dewey (1922) and some of the experiences of the residents and workers of the settlement houses in the burgeoning immigrant neighbourhoods of large US cities in the early twentieth century are instructive in this regard. Professional women such as Jane Addams, who established Hull House in Chicago, took their ideas about community development and engagement with others directly into these neighbourhoods when they opened settlement houses in which they themselves lived (Addams 1968 [1910]). In the settlement houses they experienced the meeting of their professional expertise and the community norms of the surrounding neighbourhood, and part of the activity of the settlement houses was to bring the two together. It was also an attempt to put professional knowledge at the service of the inhabitants of the poorer immigrant neighbourhoods.

For Dewey cosmopolitanism came out of situation rather than mobility or transition. Rationality was an aspect of the concrete human being in its entirety (Dewey 1922; Rosenthal 2002). It was about having to live with difference rather than just moving between different contexts. In these conditions rationality was being fully lived. This was suggested in the work of Mary Parker Follett, another urban activist of the settlement movement. Follett thought that the urban neighbourhood, with its mixing of difference, was the best example of a cosmopolitan space (Follett 1965 [1918]). This form of provincialism made local people more interesting than 'cosmopolitan people' who 'sought out homogeneous social situations' (Mattson 1998: 96). Follett explained: 'Why are provincial people more interesting than cosmopolitan [people], that is, if provincial people have taken advantage of their opportunities? Because cosmopolitan people are all alike – that has been the aim of their existence and they have accomplished it' (Follett 1965 [1918]: 195, cited in Mattson 1998: 96).

It was in the urban neighbourhood, rather than the *Gemeinschaft* community or through transnational connections, that people had to face up to heterogeneity and where democratic spirit has the potential to emerge.

From the pragmatist perspective truly cosmopolitan reason is not confined to elite knowledge or separate transnational spaces (Bridge 2005). It emerges in the urban neighbourhood in which difference is a daily reality and a negotiation. The settlement house suggests an example of everyday living with difference and, in this case, the everyday encounter between the professional knowledge of the settlement workers and the practical knowledge of the neighbourhood residents. Addams (1968) is at pains to convey how the experience of living in the poor neighbourhood did not just challenge and change her personally but also challenged and changed her professional knowledge. The social practices and ways of developing knowledge and competences of the settlement workers and neighbourhood residents gives some prospect of a city that is not divided by professional knowledge and practical knowledge. It provides a counterpoint to the cosmopolitan rationality of the global district of the new middle class and the practical rationality of the surrounding working class and national middle-class 'provincial city'. It is a form of cosmopolitan rationality that comes from location, rather than transition, a cosmopolitan form of reason lived daily in the city of difference.

Conclusions

Cosmopolitan urbanism potentially plays host to forms of cosmopolitan knowledge based on transversal rationality and associated with an international professional new middle class. Yet the cultural capital that makes cosmopolitan knowledge possible often derives from strategies of localisation over its reproduction. These localised strategies, over good taste in housing or schooling for instance, are often inimical to the openness to difference that cosmopolitanism implies. I have suggested this through the examples of hotspots of cultural capital (in the gentrification premium) and at its edges where it becomes fretted and frayed (the case of provincial gentrification). We perhaps need to recast our understanding of cosmopolitan knowledge to be one that might be provincial in its scope, but that comes out of a daily living with difference; to be one that involves the disruption of expert knowledge by a daily accommodation with competing practical knowledges; to a cosmopolitan urbanism based not on transversality and mobility, but on transaction and situation.

References

Addams, J. (1968 [1910]) *Twenty Years at Hull House*, New York: Macmillan.
Atkinson, R. and Bridge, G. (eds) (2005) *Gentrification in a Global Context: The New Urban Colonialism*, London: Routledge.

Ball, M., Bowe, R. and Gerwitz, S. (1995) 'Circuits of schooling: a sociological explanation of parental choice of school in social class contexts', *Sociological Review*, 42: 53–78.

Bourdieu, P. (1984) *Distinction: A Social Critique of the Judgement of Taste*, London: Routledge and Kegan Paul.

Bridge, G. (2001a) 'Estate agents as interpreters of economic and cultural capital: the gentrification premium in the Sydney housing market', *International Journal of Urban and Regional Research*, 25: 87–101.

Bridge, G. (2001b) 'Bourdieu, rational action and the time-space strategy of gentrification', *Transactions of the Institute of British Geographers*, NS 26: 205–16.

Bridge, G. (2003) 'Time-space trajectories in provincial gentrification', *Urban Studies*, 40: 2545–56.

Bridge, G. (2005) *Reason in the City of Difference: Pragmatism, Communicative Action and Contemporary Urbanism*, London: Routledge.

Butler, T. (1997) *Gentrification and the Middle Classes*, Aldershot: Ashgate.

Butler, T. and Robson, G. (2001) 'Social capital, gentrification, and neighbourhood change in London: a comparison of three south London neighbourhoods', *Urban Studies*, 38: 2145–62.

Butler, T. and Robson, G. (2004) *London Calling: The Middle Classes and the Remaking of Inner London*, Oxford: Berg.

Castells, M. (1996) *The Rise of the Network Society*, vol. 1: *The Information Age: Economy, Society and Culture*, Oxford: Blackwell.

Conley, V. A. (2002) 'Chaosmopolis', *Theory, Culture and Society*, 19: 127–38.

Dewey, J. (1922) *Human Nature and Conduct*, New York: Henry Holt.

Dharwadker, V. (2001) *Cosmopolitan Geographies: New Locations in Literature and Culture*, London: Routledge.

Ehrenreich, B. and Ehrenreich, S. (1979) 'The professional-managerial class', in P. Walker (ed.), *Between Capital and Labour,* Sussex: Harvester Press.

Follett, M. P. (1965 [1918]) *The New State: Group Organisation and the Solution of Popular Government*, Gloucester, MA: Peter Smith.

Gouldner, A. (1979) *The Future of Intellectuals and the Rise of the New Class: A Frame of Reference, Theses, Conjectures, Argumentation and a Historical Perspective*, London: Macmillan.

Habermas, J. (1984) *The Theory of Communicative Action*, vol. 1: *Reason and the Rationalisation of Society,* trans. T. McCarthy, London: Heinemann.

Habermas, J. (1987) *The Theory of Communicative Action*, vol. 2: *A Critique of Functionalist Reason*, trans. T. McCarthy, Cambridge: Polity.

Hage, G. (1997) 'At home in the entrails of the west', in H. Grace, G. Hage, L. Johnson, J. Langsworth and M. Symonds (eds), *Home/World: Space, Community and Marginality in Sydney's West*, Sydney: Pluto.

Hannerz, U. (1990) 'Cosmopolitans and locals in a world culture', *Theory, Culture and Society,* 7: 237–51.

Hannerz, U. (1996) *Transnational Connections: Culture, People, Places*, London: Routledge.

Hillier, J. and Rooksby, E. (2002) *Habitus: A Sense of Place*, Aldershot: Ashgate.

Isin, E. (ed.) (2000) *Democracy, Citizenship and the Global Economy*, London: Routledge.

Jager, M. (1986) 'Class definition and the aesthetics of gentrification: Victoriana in Melbourne, Australia', in N. Smith and P. Williams (eds), *Gentrification of the City*, London: Allen and Unwin.

Ley, D. (1996) *The New Middle Class and the Remaking of the Central City*, Oxford: Oxford University Press.

Ley, D. (2004) 'Transnational spaces and everyday lives', *Transactions of the Institute of British Geographers*, NS 29: 151–64.

Mattson, K. (1998) *Creating a Democratic Public: The Struggle for Urban Participatory Democracy During the Progressive Era*, University Park, PA: Penn State Press.

May, J. (1996) '"A little taste of something more exotic": the imaginative geographies of everyday life', *Geography*, 81: 57–64.

Mills, C. (1988) '"Life on the upslope": the postmodern landscape of gentrification', *Environment and Planning D: Society and Space*, 6: 169–89.

Podmore, J. (1998) '(Re)reading the "loft living" habitus in Montreal's inner city', *International Journal of Urban and Regional Research*, 22: 283–302.

Robbins, B. (2001) 'The village of the liberal managerial class', in V. Dharwadker (ed.), *Cosmopolitan Geographies: New Locations in Literature and Culture*, London: Routledge.

Rose, D. (1984) 'Rethinking gentrification: beyond the uneven development of Marxist theory', *Environment and Planning D: Society and Space*, 2: 47–74.

Rosenthal, S. (2002) 'Habermas, Dewey and the democratic self', in M. Aboulafia, M. Bookman and C. Kemp (eds), *Habermas and Pragmatism*, London: Routledge.

Sassen, S. (2000) *The Global City: New York, London and Tokyo*, Princeton, NJ: Princeton University Press.

Schrag, C. O. (1992) *The Resources of Rationality: A Response to the Postmodern Challenge*, Bloomington, IN: Indiana University Press.

Welsch, W. (1995) 'Transculturality – the puzzling form of cultures today', in M. Featherstone and S. Lash (eds), *Spaces of Culture: City, Nation, World*, London: Sage.

Welsch, W. (1998) 'Rationality and reason today', in D. Gordon and J. Niznik (eds), *Criticism and Defence of Rationality in Contemporary Philosophy*, Amsterdam and Atlanta, GA: Rodopi.

Zukin, S. (1991) *Landscapes of Power: From Detroit to Disney World*, Berkeley, CA: University of California Press.

4 Strangers in the cosmopolis

Kurt Iveson

Visions of cosmopolis

> The great possibility of the mongrel cities of the 21st century is the dream of *cosmopolis*: cities in which there is acceptance of, connection with, and respect and space for 'the stranger', the possibility of working together on matters of common destiny and forging new hybrid cultures and urban projects and ways of living.
>
> (Sandercock 2003: 127)

In this chapter, I pay some critical attention to visions of the city as *cosmopolis*. Influenced by, and contributing to, the revival of cosmopolitanism in social and political theory more generally (see, for example, Cheah and Robbins 1998; Archibugi 2003), these visions of cosmopolis are intended to act as a counter-factual against which the conditions of urban life in particular places might be critically compared. For Sandercock (1998: 7), the cosmopolis is not an actually existing city, but rather a 'construction site of the mind', a kind of social imaginary designed to offer both grounds for critique and inspiration for alternatives. Similarly, Tajbakhsh (2001: xv) argues that 'the idea of cosmopolis is a radical alternative to both social homogeneity and a plurality of mutually exclusive enclaves – which in the end are the same'. The influence of cosmopolitanism in contemporary urban theory should come as no surprise – the city has always been a privileged site for those interested in cosmopolitan ethics and politics, as a space in which parochial loyalties to kin, tribe, 'race' or nation might give way to more radically uncertain and explicitly humanitarian identifications.

While these visions of cosmopolis are diverse in their scope and shape, one important figure appears time and time again – *the stranger*. In focusing on the stranger, cosmopolitan urban theorists are turning to a figure which has a distinguished career in writings on urban life. Indeed, for some writers the co-presence of strangers literally defines modern urban life. To offer but a few examples:

- 'To live in a city is to live in a community of people who are strangers to each other' (Raban 1974: 15).
- 'City life is a being together of strangers' (Young 1990: 240).
- 'The city brings together people who are different, it intensifies the complexity of social life, it presents people to each other as strangers' (Sennett 1994: 25–6).
- 'City life is carried on by strangers among strangers' (Bauman 1995: 126).

If city life is in essence 'lived among strangers', then attempts to order urban life which embody 'enclave consciousness' (Tajbakhsh 2001), 'fear of touching' (Sennett 1994), fear of 'the other' (Sandercock 1998), or the desire for 'community' (Donald 1999) are deemed to be inherently problematic. In contemporary cities, such trends are said to be all too evident, in the form of, for example: gated communities and other forms of class and/or ethnic residential segregation; privately owned public spaces with security measures designed to identify and remove 'troublemakers'; the widespread introduction of new surveillance technologies such as closed-circuit television and police helicopters; urban design measures and punitive statutes designed to reduce crime and so-called 'anti-social behaviour'; xenophobic mobilisations against asylum seekers, refugees and immigrants. Such efforts to protect place-bound identities through the construction of territorial boundaries are considered no more than doomed and yet dangerous attempts to forge a 'hegemonic closure' (Tajbakhsh 2001) which fixes and sustains identities ('us' and 'them') in the face of the inherent diversity and/or hybridity of urban identities. The consequences are profound. For Deutsche (1999: 176), 'the fact that members of certain social groups are treated as detestable strangers with no "right to the city" . . . and are, as a consequence, threatened in the city by its "rightful" citizens, places the city itself in peril'. Sandercock (2002: 216–17) is even more direct – for her, initiatives informed by a fear of strangers and others literally threaten to 'kill the city'.

But who or what is a 'stranger'? Given the importance of the stranger in cosmopolitan urban theory, in this chapter I will look at the different answers to this question as a means to critically engage with the cosmopolitan urban social imaginary. Across this literature, two distinct understandings of strangers in urban life are most commonly put to work in developing visions of cosmopolis. One understanding considers 'the stranger' as a particular kind of body from elsewhere. The other understanding treats 'estrangement' as a condition of urban life. In what follows, I elaborate upon these two understandings of strangers/strangeness and urban life, showing how they inform visions of cosmopolis in the existing literature. My purpose here is not to divide the cosmopolitan literature into two distinct 'camps' (indeed, the work of individual authors sometimes mobilises both of these approaches for different purposes). Rather, my aim is to identify some of the strengths and limitations of visions of cosmopolis which are informed by either or both of these two understandings of strangers in urban life.

'The stranger' and the hospitable cosmopolis

Simmel's observations on the nature of modern urban life at the turn of the twentieth century included a series of influential reflections on 'the stranger'. He considered the figure of the stranger to embody a 'unity of nearness and remoteness'. The stranger is not:

> the wanderer who comes today and goes tomorrow, but rather . . . the person who comes today and stays tomorrow . . . He is fixed within a particular spatial group, or within a group whose boundaries are similar to spatial boundaries. But his position in this group is determined, essentially, by the fact that he has not belonged to it from the beginning, that he imports qualities into it, which do not and cannot stem from the group itself.
>
> (Simmel 1950 [1903]: 402)

For Simmel, then, the stranger is the product of an *arrival* which has both spatial and temporal dimensions. The stranger has moved from one place to another, at some moment passing across the boundaries of a pre-existing group with the hope of staying. This arrival brings strangers into contact with other individuals to whom they are not connected through ties of kinship, locality and occupation (Simmel 1950 [1903]: 404).

For some, this focus on arrivals and their consequences for urban life gives 'the stranger' on-going analytical value for investigations into contemporary urban life. While Simmel was primarily concerned with consequences of massive rural–urban migration for European cities at the turn of the twentieth century, some contemporary theorists are preoccupied with the consequences of international movements of people for cities of the twenty-first century. Sandercock (2003), for example, argues that global flows of people and culture have produced new demographic realities for Western cities. 'Strangers' continue to arrive from elsewhere in even greater numbers, and their 'integration' presents a series of problems for urbanites and urban theory. The arrival of strangers is perceived as a threat to an existing 'socio-spatial and socio-temporal sense of place and identity', a disruption to 'taken-for-granted categories of social life and urban space' (Sandercock 2003: 110, 134).

For those cosmopolitan theorists who understand the stranger in this way, a key question is: how should 'locals' respond to the disruption caused by the arrival of 'the stranger'? Some notion of *hospitality* becomes central to the construction of a true cosmopolis. Kant's discussion of cosmopolitan hospitality remains influential. He defined hospitality as 'the right of a stranger not to be treated with hostility when he arrives on someone else's territory' (Kant, quoted in Derrida 2000: 5). He did not understand this to be a matter of the absolute flattening of the Earth, through the dissolution of all borders, boundaries and territorial identities. Rather, Kant argued in favour of making those borders and boundaries more porous, insisting on some right to hospitality which is extended to those who do not belong as full citizens of a given nation-state (Dikeç 2002).

Kahn (1987) provides us with a contemporary example of this Kantian approach in urban studies. According to Kahn (1987: 12), 'Cities do not create strangers. Cities are the refuge of strangers. Cities are the place for men and women already estranged. And cities are to be judged by their welcome.' She refuses to blame cities for disrupting pre-existing modes of sociality by bringing strangers with different values into proximity. Indeed, this is an inherent part of urban life, so cities in fact have an obligation to extend a welcome to strangers. In this reading, the stranger is an outsider who has arrived from elsewhere, and is on someone else's territory. The stranger's arrival can be welcomed (or not). For the cosmopolitan, then, the good city ought to welcome the stranger, avoiding any temptation to lapse into 'stranger danger' by treating the stranger as a threat to be excluded. The construction of territorial boundaries designed to purge the city of strangers is considered both unjust and ultimately unrealistic. Planning approaches 'which seek to banish fear by either banishing or transforming those seen as inducing fear, have surely exhausted themselves in their futility' (Sandercock 2003: 109).

Visions of cosmopolis which promote hospitality as a progressive response to the arrival of 'the stranger' raise the question: who will set the terms of the 'welcome' extended to strangers? Here, Simmel's 'pre-existing group' potentially returns to haunt this particular cosmopolitan frame. If the locals/natives choose to be hospitable, are they not still distinguishing themselves from the strangers on territorial grounds? The notion of hospitality implies that there is some place (in this instance, the city) *into which* the stranger should be welcomed, and *over which* the host has mastery to determine the conditions of the welcome. As Derrida (2000: 14) puts it, 'It does not seem to me that I am able to open up or offer hospitality, however generous, even in order to be generous, without reaffirming: this is mine, I am at home, you are welcome in my home.'

Studies into the operation of multiculturalism in a variety of national contexts illustrate the potential for some forms of hospitality to reinforce, rather than challenge, political inequalities. In Denmark, Diken (1998: 53) has argued that both supporters and critics of difference multiculturalism 'speak in the same terminology, on the same object, with the same concepts, around the same table'. The categories of 'us' (the Danes) and 'them' (the non-Danes) are left intact, if configured differently in relation to one another in debates over immigration and settlement policy. Similarly, Hage (1998: 201) argues that the cosmopolite who claims the status of host in Australian debates on multiculturalism 'is a *class* figure *and* a white person, capable of appreciating and consuming "high-quality" commodities and cultures, including "ethnic" culture'. The cosmopolite's politics is underpinned with the fantasy that the Australian national space is theirs to control. In the British context, Sandercock is critical of the importance attached to 'community cohesion' in immigration and urban policies. This policy frame is used to assert the importance of shared values while implicitly defining the 'problem' to be addressed as 'a problem of "them" adjusting to "us", being gracious guests in "our home"' (Sandercock 2003: 91). For her, the problems with this state-directed vision of hospitality are neatly summed up in former British Home Secretary David Blunkett's argument that 'We have norms of acceptability and those who come into our home – for that is what it is – should accept those norms' (quoted in Sandercock 2003: 91).

Of course, such outcomes do not exhaust the practical possibilities for politics informed by notions of hospitality. My point here is that the construction of hospitality as a cosmopolitan virtue must at least confront a set of political tensions concerning the figuration of some bodies as 'hosts' and others as 'stranger'. Without an interrogation of how the designations 'host' and 'strangers' are figured, visions of cosmopolis will share key assumptions about 'the stranger' with those who are explicitly hostile to the presence of strangers (see Ahmed 2000: 4). While cosmopolitans like Kahn are critical of any boundary-drawing efforts in response to the arrival of the stranger, the very figure of 'the stranger' is the cause *and also the effect* of boundary-drawing efforts conducted by those who have assumed the status of 'host' or 'native' (Bauman 1988–9: 8).

So, to the extent that calls for hospitality are informed by an understanding of 'the stranger' as a particular body, and 'the city' as a coherent subject which can extend a 'welcome' (or not), these visions of the hospitable cosmopolis risk naturalising and privileging very particular interests over others. A more critical approach to hospitality would need to bring the very subjecthood of 'the city' into question, by troubling the claims of those who seek to *welcome* strangers as well as contesting those who would seek to restrict or condition the entry of strangers. I now turn to another way of understanding strangers and urban life which is used to sustain such alternative visions of cosmopolis.

'Estrangement' and the cosmopolis as a 'being together of strangers'

In his work on cosmopolitanism, Derrida (2001: 7–8) wondered whether 'the City' might have an important role to play in addressing the injustices of the international system of nation-states which is increasingly inhospitable to those who are 'without papers': 'Could the City, equipped with new rights and greater sovereignty, open up new horizons of possibility previously undreamt of by international state law?' For this to be so, a fundamentally different conception of 'the City' is required – one which does not take its subjecthood for granted. The journey towards cosmopolis ought not involve the restoration of:

> an essentially classical concept of the city by giving it new attributes and powers; neither would it be simply a matter of endowing the old subject we call 'the city' with new predicates. No, we are dreaming of another concept, of another set of rights for the city, of another politics of the city.
>
> (Derrida 2001: 8)

In pursuit of such alternative concepts of the city, some writers have mobilised a different understanding of strangers and urban life. Pushing past a cosmopolitanism underpinned by the designation of some bodies as hosts and others as strangers, they have instead sought to explore the possibilities for ethical and political engagements which challenge such categorisations (Dikeç 2002). Central to these efforts is the notion that *strangeness is a condition* shared by everybody rather than a property of some-bodies. This universal estrangement is thought to be an inevitable product of (post-)modern urban life, which produces countless displacements for its inhabitants, such that the notion of a purified 'home ground' for any individual is ultimately untenable (Bauman 1988–9: 39). Urban inhabitants move through different neighbourhoods, they participate in a range of functionally and geographically differentiated 'life spaces' that constitute the routines of urban life (work, education, entertainment, consumption, sociability and so on) (see Bech 1998: 216–17). Consequently, 'There is no hiding place. There is nowhere you can go and only be with people who are like you' (Reagon 1981, cited in Sandercock 2003: 127).

This understanding of strangers shares Simmel's emphasis on mobility, but the concept of mobility is radically extended. Every individual is a 'partial stranger' because arrivals (and departures) are incessant, with these displacements calling forth a never-ending series of responses and adjustments. Indeed, this revised understanding of strangeness is attractive for many contemporary urban and social theorists precisely because of its emphasis on individual mobilities associated with immigration, urbanisation, tourism, etc. (Diken 1998: 125). Of course, this is not to say that urban dwellers necessarily share similar *experiences* of increasing mobility. Indeed, urban populations are in some ways stratified by their different mobilities (Bauman 1995: 130; Pels 1999: 76). The point for cosmopolitan theorists is not so much that individuals share actual experiences of mobility, but rather that they share a *condition* of hybridity, of indeterminacy – that is, a condition of estrangement. As Dillon (1999: 95, cited in Dikeç 2002: 240) puts it: 'Estrangement is . . . a condition of human existence as a *how*, a way of being, and *not a what*, an object whose essence may be captured in a concept.'

What are the implications of this revised understanding of estrangement for visions of cosmopolis? For those who share this view of urban life, the key question for urban theory shifts from the question of how 'locals' should respond to 'the stranger' to the question of how all urban inhabitants should respond to their mutual estrangement.

Cosmopolitan urban theorists have identified and criticised a range of common responses to this estrangement. The desire for shared community values in the face of estrangement is criticised on the grounds that it tends to promote the very boundary-drawing efforts that produce the fortress city. Communitarian visions of city life are in the end a 'longing for harmony among persons, for consensus and mutual understanding' (Young 1990: 229). This longing inevitably produces boundary-drawing efforts which seek to distinguish those who belong from those who do not (Donald 1999: 154). Mutual indifference, or indifference to difference, is another way of living as strangers that is feared to be all too common in many cities. Some cosmopolitans share Sennett's (1994: 357) worry that 'the sheer fact of diversity does not prompt people to interact'. In such circumstances, he wonders 'whether a civic culture can be forged out of human differences'. Scapegoating of strangers is also criticised. This might involve the construction of some-bodies as strangers, with a further categorisation of strangers which imbues some with the positive values associated with strangeness (the sexy, interesting, desirable strangers), while others are 'selected out to condense and concretise the undesirable sides of the stranger-in-general as unambivalent images' (Diken 1998: 135).

What would a cosmopolitan response to estrangement and ambivalence look like? Kristeva's (1991) *Strangers to Ourselves* has been particularly influential (see, for example, Deutsche 1999; Donald 1999; Sandercock 1998, 2003; Tajbakhsh 2001). In the following frequently cited passage, she presents an understanding of 'the foreigner' as immanent to our being:

> The foreigner is within me, hence we are all foreigners. If am a foreigner, there are no foreigners . . . The ethics of psychoanalysis implies a politics: it would involve a cosmopolitanism of a new sort that, cutting across governments, economies and markets, might work for a mankind whose solidarity is founded on the consciousness of its unconscious.
>
> (Kristeva 1991: 192)

Life among strangers in the city, then, should be a matter of opening up to the *hybridity* and instability of our own identities, thus putting them at risk. In Tajbakhsh's (2001: 183) terms, the 'promise of the city' is not so much the promise of hospitality towards strangers, or the peaceful and productive mingling of different groups, as 'the freedom to glimpse our own hybridity, our own contingency'. This can only emerge from a 'cosmopolitan ethic' of 'openness to the other proceeding from the recognition of the stranger within us' (Tajbakhsh 2001: 9). Similarly, for Donald (1999: 170), the capacities required to participate in dialogue with strangers 'shade into and out of the virtues made possible by the great adventure of the city: politeness as well as politics, civility as well as citizenship, the stoicism of urbanity, the creative openness of cosmopolitanism'. The implication here is that if we are all strangers, and there are no hosts, then there can be no set of values or standards for living together in the city which are beyond contestation.

Exactly what might this 'openness' to others involve? Some of the literature remains rather vague on this point, particularly in its resort to a kind of universal communicative ethics. Both Young (2000) and Deutsche (1999) propose that dialogue among urban inhabitants who share a condition of mutual estrangement is a matter of being '*reasonable*'. Being reasonable means being open to surprise, change and indeterminacy in relations with ourselves and others, thereby refusing to fix one's own identity, or the identities of others, in advance of encounters with them. It is to avoid retreats into external sources of meaning and value. However, some have argued that urban theory must not be content to construct such ideals of virtuous citizenship. Donald urges us to adopt a focus on 'thicker description of the openness to unassimilable difference, and so also a concern with the mundane, pragmatic but sometimes life-or-death arts of living in the city' (1999: 170). One of the strengths of some cosmopolitan urban theory – and one of its potentially significant contributions to theories of cosmopolitan democracy more generally – is its attention to the details of ethical and political *labours* which are required to construct any kind of togetherness in difference. Most notably, Sandercock's (1998, 2002, 2003) various writings on cosmopolis and insurgent planning are informed by investigations into the conduct of urban politics and planning in a variety of contexts. Of particular interest to her in her most recent work are the:

> micro-publics where dialogue and prosaic negotiations are compulsory, in sites such as the workplace, schools, colleges, youth centres, sports clubs, community centres, neighbourhood houses, and the micro-publics of 'banal transgression' (such as colleges of further education), in which people from different cultural backgrounds are thrown together in new settings which disrupt familiar patterns and create the possibility of initiating new attachments . . . Part of what happens in such everyday contacts is the overcoming of feelings of strangeness in the simple process of sharing tasks and comparing ways of doing things.
>
> (Sandercock 2003: 94)

The purpose of Sandercock's vision of cosmopolis is precisely to inform and inspire the exploration of these possibilities.

But even in the work of those who are keen to explore *how* common interests might be made through empirical investigation, there is sometimes a tendency to fall back on problematic ideals of cosmopolitan virtue when it comes to considering *why* urban inhabitants ought to get involved in such efforts. In the absence of uncontested values, it is commonly argued that the shared problems of being together in the city necessitate some kind of dialogue across heterogeneity and hybridity. In the cosmopolis, urban inhabitants must acknowledge their mutual *fates* as well as their mutual *estrangement*. The cosmopolis is not envisaged as a city characterised by its hospitality towards 'the stranger', but rather as a city which facilitates the agonistic and democratic 'being together of strangers' (Young 1990). Living together in the city, these strangers must confront shared problems, even if they do not share identities.

This claim about the shared problems of urban life is often mobilised to ground normative critiques of those who fail to adjust to the 'demands' of contemporary urban life in the prescribed fashion. For Tajbakhsh (2001: 6, emphasis added), in the democratic city there is a 'moral *demand* placed on all those who share a common territory and polity to enter into a public realm with some sense of unifying values and beliefs, however tentative and provisional'. Sandercock (1998) contends that while our differences should be encouraged to flourish in the cosmopolis, we must also 'acknowledge some sense of belonging to a broader polity and society': 'This does not necessarily mean the return of the outmoded concept of "the public interest", but it does *demand* the creation of a civic culture from the interactions of multiple publics' (1998: 186–7, emphasis added). She attacks the 'new urban socio-spatial segregations' on the grounds that they 'increase the difficulty of engaging a variety of social groups in political life, in which common goals and solutions would have to be negotiated' (2003: 125). Living together in the city, then, '*demands* a response in terms of ethics as well as politics' (Donald 1999: 170, emphasis added).

The mobilisation of this normative ideal of cosmopolis to make demands of urban inhabitants is either explicitly or implicitly directed towards those powerful groups who seek to enclose themselves in bounded spaces of privilege, thereby failing to acknowledge their mutual fate with others in the city. But the foundation of this demand in a vision of the city as a 'being together of strangers' has wider implications which are more troubling. To claim that strangeness is a condition of being together in the city, a *how* not a *what*, is to make an ontological claim that in the city 'we are all strangers'.[1] Sara Ahmed (2000: 73) worries that this ontology of strangers 'perform[s] the gesture of killing the strangers it creates, by rendering them a universal: a new community of the "we" is implicitly created – "we" are a community of strangers'. This vision of a community of strangers, like other visions of community, is sustained through a drawing of boundaries which appear to be outside of politics.

The nature of this boundary is worth considering in further detail. When the cosmopolis is understood as a 'being together of strangers', the common ground of the 'community of strangers' is defined by the boundary of 'the city' itself. This need not be considered as a line on a map which defines the limits of the city. Rather, the boundaries of the city are explicitly drawn between those who are willing to acknowledge the essential common ground of their togetherness and those who are not. Those who will not acknowledge their togetherness are not only seeking to escape the urban, they are also threatening its very existence. They are anti-urban because they are unwilling to acknowledge their common ground as a community of strangers. Importantly, this cosmopolitan specification of 'common ground' is of a fundamentally different order to the common ground specified by communitarians, who place some eternal set of 'shared community values' outside of the realm of contestatory politics. Deutsche (1999: 184–5) makes this distinction clear: 'Because the ground of our commonality is uncertain, has no fixed point, no external guarantee, it is open to interrogation through the declaration of rights, a process in which the meaning and unity of society are perpetually negotiated.' Any solutions which emerge from such negotiations are only ever partial or temporary. Nonetheless, we ought to be wary of the cosmopolitan insistence that urban living *demands* an orientation to address shared problems with strange others. Such an insistence is likely to privilege a particular ethical orientation (such as, say, the capacity to be 'reasonable') as a universal cosmopolitan virtue.

The risk here is that some model of the virtuous cosmopolitan citizen – no longer bound to others by kin, tribe or even nation, but bound instead by an elective commitment to living together as strangers – could just as easily be mobilised to *police* the city as well as to inspire a *politics* of the city.[2] So, for example, if graffiti writers refuse to put their identities at risk by engaging in a wider dialogue over urban aesthetics with 'outsiders' to their subculture, are they being unreasonable? Should they be lured or forced into such a dialogue, because living together in the city demands it? If the homeless and the addicted fail to participate in a debate about the norms which govern street contacts and begging, are they being unreasonable? Should they be 'educated' or 'empowered' to participate in such a debate, because living together in the city demands it? If it is not the police who intervene to remove and punish these 'unreasonable' people, then it is the youth worker, the local government official, or the educator who might step in to police the good city by providing opportunities designed to impart the proper capacities and skills. Perhaps not coincidently, the capacities which such a model of cosmopolitan urbanism would render virtuous are a reflection of the capacities of those professionals (architects, planners, educators, academics, etc.) who sometimes promote them (Isin 2002). Any demand that all urban inhabitants adopt a cosmopolitan openness to others will have a fundamentally different meaning for weak groups than it will for those who have voluntarily fortified themselves in enclaves of privilege. In some instances, the construction of fragile boundaries which limit engagements with others is all that protects the weak from annihilation – some boundaries and exclusions might well be politically justified (Iveson 2003). Indeed, bounded spaces may constitute sites of resistance against hegemonic identities in the city (see Watson and Gibson 1995: 257).

Cosmopolitanism and 'the city'

My purpose in this chapter has been to offer some critical reflections on visions of cosmopolis as alternative urban social imaginaries, by investigating the ways in which the stranger has been figured by cosmopolitan urban theorists. I have argued that in the work of these theorists distinct understandings of the stranger and urban life have been put to work in constructing cosmopolitan visions of the good city as a 'refuge of strangers' and/or a 'being together of strangers'. These visions of the good city are mobilised to provide a normative justification for claims that urban inhabitants ought to (i) welcome strangers from elsewhere, and/or (ii) commit to participation in an urban civic culture which is premised on a sense of shared fates and mutual estrangement rather than a sense of shared values. The dream of cosmopolis is the dream of a city where the actions of planners and citizens are informed by a cosmopolitan ethic of hospitality and/or openness.

So, while we can distinguish between two distinct ways in which cosmopolitan urban theorists understand the figure of the stranger, these distinct understandings are mobilised in quite similar ways. The call for urbanites to embrace cosmopolitan structures of feeling is premised on the claim that such structures of feeling are ethical responses to the nature of urban life. As Conley (2002: 130) puts it, 'cosmopolites, from their city, think towards the horizon'. For her (2002: 130), 'it is not enough to denounce an isolated incident such as this example of child labor, or child soldiers, or of sweatshops'.

The implication here is that a whole range of exclusionary actions in cities can be understood and opposed as instances of a more universal hostility towards the 'stranger' in 'the city'. So, for example, hostility towards asylum seekers in European cities is condemned because it represents a failure to welcome the stranger. The fortification of city centre spaces against groups of teenage youths on behalf of 'the community' is condemned because it represents a failure to acknowledge mutual estrangement and responds to a lack of shared values with fear rather than openness. The problem with this critical strategy is that it seeks to ground urban cosmopolitanisms in universal models of city life and identity which stand outside of time and place. The construction of alliances across different and larger cosmopolitan structures of feeling therefore appears as the product of a cosmopolitan mindset towards the stranger/estrangement, rather than the product of hard political slog. But as Bruce Robbins (1998: 6) notes:

> Larger loyalties can either be there or not there. They have to be built up laboriously out of the imperfect historical materials . . . that are already at hand. They do not stand outside of history like an ultimate court of appeal.

In this context, I think the most promising direction set out in some of the cosmopolitan urban theory I have considered is the call for urban researchers to focus on the messy, difficult and risky labours which are required for the construction of any sense of solidarity or alliance which transcends territorially fixed identity categories. But I am not convinced that the figure of the 'stranger' is useful for thinking about these labours. Like Laurier *et al.* (2002: 353, emphasis in original), I remain suspicious of the way in which urban theory posits 'the type *stranger* as if it were used as ubiquitously and reductively in ordinary interactions as it is by social theorists'. If these ordinary interactions must provide the 'imperfect historical materials' out of which hybrid identifications might be built, surely critical urban theory needs to be more attuned to the *specifics* of these interactions? As Laurier *et al.* (2002: 353) go on to argue:

> The massively apparent fact is that people in cities do talk to one another as customers and shopkeepers, passengers and cabdrivers, members of a bus queue, regulars at cafes and bars, tourists and locals, beggars and by-passers, Celtic fans, smokers looking for a light, and . . . as neighbours.

In thinking about the laborious construction of larger political loyalties as the efforts of people who are 'strangers' to each other and themselves, theorists of the cosmopolis are in danger of obscuring the quite particular identifications which are central to the ethical and political projects they hope to inspire. An understanding of urban inhabitants as 'strangers' who must live together rather than 'neighbours' or 'workers' (or whatever) posits a basis for their togetherness which is in effect ontological rather than political.

Rather than making assertions about the proper basis of common interest in the cosmopolis, urban theorists might more usefully ask *how* and *why* structures of feeling which identify 'shared problems' are produced in *specific circumstances*. For example, in her examination of the formation of alliances between women and foreign 'Others' in London in the early twentieth century, Mica Nava (2002) traces the emergence of a specific cosmopolitan and anti-racist imagination to a 'dynamic interconnection between identification and distantiation – between desire and repudiation' (2002: 88). Openness to foreigners from British colonies was an expression of women's 'sense of psychosocial dislocation and non-belonging' to English culture which oppressed them rather than the product of an ethical or intellectual critique of nationalism (2002: 89). Thinking about the construction of less overtly 'political' common interests/problems, the study by Laurier *et al.* (2002) of the work done by neighbours in making the loss of a cat a shared problem also focuses on the particular circumstances and identifications that are mobilised. The point in both of these examples is that alliances and common interests are not premised on a mutual recognition of others as 'strangers'. Rather, alliances and common interests are made out of quite specific historical and geographical materials – there is no basis for these alliances and common interests in some model of 'city life' that stands outside of place and time. As Ahmed puts it:

> Alliances are not guaranteed by the pre-existing form of a social group or community, where that form is understood as commonality (a community

> of friends) or uncommonality (a community of strangers). The collective then is not simply about what 'we' have in common – or what 'we' do not have in common. Collectives are formed through the very work that we need to do in order to get closer to others.
>
> (Ahmed 2000: 179–80)

In conclusion, then, I think that a consideration of how the figure of 'the stranger' is understood and put to work in cosmopolitan urban theory points to some important problems for visions of cosmopolis. In seeking to inspire some notion of 'the city' as a common ground through which hybrid identifications and alliances will take shape, theorists of cosmopolis tend to mobilise normative models of city life in order to justify calls for hospitality and openness. By doing so, they seem to imply that some model of 'the city' as cosmopolis might serve as Robbins' 'ultimate court of appeal'. But the question of whether 'the city' might be a useful political resource for those engaged in attempts to construct cosmopolitan ethical and political horizons is not one that can be settled in advance of politics. 'The city' might be useful, and it might not. The grounds for critique of some of the more pernicious aspects of contemporary urbanisation – such as gated communities or reactionary mobilisations against asylum seekers – might just have to be *made* through the very labour which attempts to address these particular issues. I think there is good reason to be wary of the notion that progressive urban politics should be derived with reference to a theoretical model of the 'good city' as cosmopolis which renders gated communities or hostility towards asylum seekers as but particular instances of hostility towards 'the stranger' in urban life.

Notes

1 Some writers, it should be noted, tend to move between strangeness as a 'how' and a 'what'. For instance, at various points Sandercock (2003) stresses the need for respect for and connection with 'the stranger' as a particular body (most often, but not always, the 'immigrant'), while at other times she speaks of 'all of us' as strangers. In moving between these figurations of the stranger/strangers, she argues that the construction of some bodies as strange is the product of political economic processes, while the key to undoing such constructions lies in an acknowledgement of a shared condition of strangeness informed by social-psychological readings of individual subjectivity (see 2003: 123).

2 On the distinction between politics and police, see Rancière (1999).

References

Ahmed, S. (2000) *Strange Encounters: Embodied Others in Post-Coloniality*, London: Routledge.

Archibugi, D. (ed.) (2003) *Debating Cosmopolitics*, London: Verso.

Bauman, Z. (1988–9) 'Strangers: the social construction of universality and particularity', *Telos*, 78: 7–42.

Bauman, Z. (1995) *Life in Fragments: Essays in Postmodern Morality*, Oxford: Blackwell.

Bech, H. (1998) 'Citysex: representing lust in public', *Theory, Culture and Society*, 15: 215–41.

Cheah, P. and Robbins, B. (eds) (1998) *Cosmopolitics: Thinking and Feeling Beyond the Nation*, Minneapolis, MN: University of Minnesota Press.

Conley, V. A. (2002) 'Chaosmopolis', *Theory, Culture and Society*, 19: 127–38.

Derrida, J. (2000) 'Hospitality', *Angelaki*, 5: 3–18.

Derrida, J. (2001) *On Cosmopolitanism and Forgiveness*, London: Routledge.

Deutsche, R. (1999) 'Reasonable urbanism', in J. Copjec and M. Sorkin (eds), *Giving Ground: The Politics of Propinquity*, London: Verso.

Dikeç, M. (2002) 'Pera peras poros: longing for spaces of hospitality', *Theory, Culture and Society*, 19: 227–47.

Diken, B. (1998) *Strangers, Ambivalence and Social Theory*, Aldershot: Ashgate.

Donald, J. (1999) *Imagining the Modern City*, London: Athlone.

Hage, G. (1998) *White Nation: Fantasies of White Supremacy in a Multicultural Society*, Sydney: Pluto Press.

Isin, E. (2002) *Being Political: Genealogies of Citizenship*, Minneapolis, MN: University of Minneapolis Press.

Iveson, K. (2003) 'Justifying exclusion: the politics of public space and the dispute over access to McIvers ladies' baths, Sydney', *Gender, Place and Culture*, 10: 215–28.

Kahn, B. M. (1987) *Cosmopolitan Culture: The Gilt-Edged Dream of a Tolerant City*, New York: Atheneum.

Kristeva, J. (1991) *Strangers to Ourselves*, London: Harvester Wheatsheaf.

Laurier, E., Whyte, A. and Buckner, K. (2002) 'Neighbouring as an occasioned activity: "Finding a lost cat"', *Space and Culture*, 5: 346–67.

Nava, M. (2002) 'Cosmopolitan modernity: everyday imaginaries and the register of difference', *Theory, Culture and Society*, 19: 81–99.

Pels, D. (1999) 'Privileged nomads: on the strangeness of intellectuals and the intellectuality of strangers', *Theory, Culture and Society*, 16: 63–86.

Raban, J. (1974) *Soft City*, London: Hamish Hamilton.

Rancière, J. (1999) *Disagreement: Politics and Philosophy*, Minneapolis, MN: University of Minnesota Press.

Reagon, B. (1981) 'Coalition politics: turning the century', in B. Smith, *Home Girls: A Black Feminist Anthology*, New York: Kitchen Table, Women of Color Press.

Robbins, B. (1998) 'Introduction', in P. Cheah and B. Robbins (eds) *Cosmopolitics: Thinking and Feeling Beyond the Nation*, Minneapolis, MN: University of Minnesota Press.

Sandercock, L. (1998) *Towards Cosmopolis: Planning for Multicultural Cities*, Chichester: Wiley.

Sandercock, L. (2002) 'Difference, fear, habitus: a political economy of urban fears', in J. Hillier and E. Rooksby (eds), *Habitus: A Sense of Place*, Aldershot: Ashgate.

Sandercock, L. (2003) *Cosmopolis II: Mongrel Cities in the 21st Century*, London: Continuum.

Sennett, R. (1994) *Flesh and Stone: The Body and the City in Western Civilization*, London: Faber and Faber.

Simmel, G. (1950 [1903]) 'The stranger', in K. H. Wolff (ed.), *The Sociology of Georg Simmel*, New York: The Free Press.

Tajbakhsh, K. (2001) *The Promise of the City: Space, Identity and Politics in Contemporary Social Thought*, Berkeley, CA: University of California Press.

Watson, S. and Gibson, K. (1995) 'Postmodern politics and planning: a postscript', in S. Watson and K. Gibson (eds), *Postmodern Cities and Spaces*, Oxford: Blackwell.

Young, I. M. (1990) *Justice and the Politics of Difference*, Princeton, NJ: Princeton University Press.

Young, I. M. (2000) *Inclusion and Democracy*, Oxford: Oxford University Press.

PART II

Consuming the cosmopolitan city: materialities and practices

Sociality and the cosmopolitan imagination: national, cosmopolitan and local imaginaries in Auckland, New Zealand

Alan Latham

Armadillo. Karangahape Road. A Thursday evening. Paul Rennie Brown is drunk again. Okay, not really drunk. Happy. Voluble. Okay, a little more than happy and voluble. He's LOUD. Well actually – to be fair – everyone's loud. Loud in a pleasant, raucous, Chardonnay-ed, Sauvignon Blanc-ed, kind of way. All eleven of them. A group of 40 plus somethings. All single, a string of sticky, confusing, messy, divorces behind them. Solo parents with no kids, as Paul likes to call them. They are having a dinner party for no other reason than that Paul thought that they might like each other. It might be fun. Come along. Enjoy yourself. Make connections. Paul likes these New Age-ey management book phrases. It is part of his charisma, his attractiveness. It keeps things interesting. Conversation flows, although it isn't really about anything much. Nothing immediately earth shattering. Talk about business. Other evenings in similar restaurants. Lots of things. Tonight was a great night.

It was also a night like many others for Paul and his friends. Armadillo is an expensively outfitted restaurant-cum-lounge bar nearly opposite the Pink Pussy Cat massage parlour at the top of Karangahape Road, West Auckland. The seats are tanned leather, attractive and comfortable. The floor is polished hardwood. The food an inventive fusion of Mediterranean, Asian and South American cuisine. It is an impressive yet welcoming place. But why should Armadillo or indeed the socializing of Paul and friends be of any interest to us social scientists? Why should we make the effort to think about a restaurant and what is going on in there?

Figure 5.1 ***Armadillo.***

Well, the initial answer is a profoundly local one. Nothing like Armadillo or the similar kinds of places around the corner on Ponsonby Road or Jervois Road existed 20 or 30 years ago in Auckland. At that time New Zealand culture, and along with it New Zealand urban culture, was defined by its stuffiness, its lack of a vigorous evening urban economy, or even much of a daytime public culture. Visiting New Zealand in the 1970s foreign visitors were impressed by its homogeneity, its same-ness, its *conformity*. It seemed to people a place where nothing happened – where indeed things had stopped happening.

By the late 1990s when our 40 plus somethings meet, the story is quite different. If we were to follow Paul around for a week we would find Paul participating in a remarkably vigorous and pluralistic urban public culture. This is public culture that is in all sorts of ways remarkably cosmopolitan. Along with the evening in the Armadillo we would see him in Italian-style espresso bars, snug Manhattan-style drinking dens open until well after midnight, smart designer kebab joints, and more. And Paul is not particularly unique. If instead of him we were to follow Julia, a financially stretched fabric designer in her early 40s who often shares babysitting with Paul, we would see

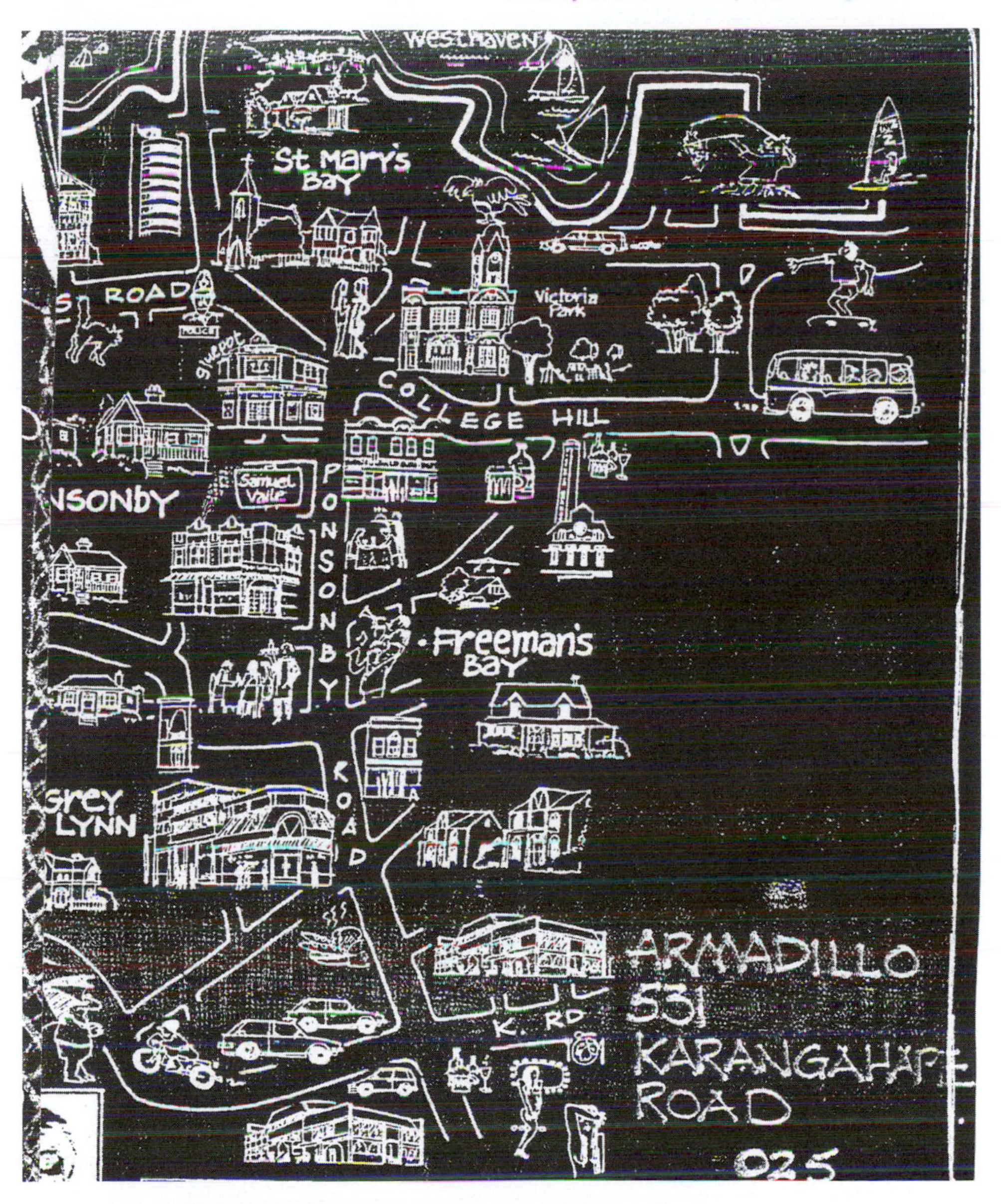

Figure 5.2 ***Karangahape Road.***

something similar. What has happened? Where did all this stuff, these ways of eating, drinking, socializing, these ways of ordering public spaces, come from? And what does it all mean for New Zealand's sense of its urban culture?

To be interested in Armadillo and the sociality of Paul and his friends is to be interested in this becoming cosmopolitan. It is to try and think about the ways that a particular urban landscape, and a particular group of people, have become seemingly more diverse, more international, more *worldly*. This apparently is to tell a story of cultural globalization. And certainly it is in part about this. But there is another reason why the presence of Armadillo and the whole new universe of cosmopolitan, global stuff that is suddenly so pervasive around inner-city Auckland and elsewhere is of interest. And this reason has a more obviously general feel to it. To explore the ways that the 'cosmopolitan' has become part of the worlds of people like Paul and Julia, the way it has become part of the organizing fabric of Auckland's public landscape, is also to think about the ways that culture is enacted. It is to think about the ways that all sorts of materials and practices are put to use and get bound up with the making of a particular culture in ways that often seem strangely alien to their original source.

In this chapter I want to explore something of what it means to think about the world becoming more cosmopolitan, and also to think about this process of putting certain kinds of material 'to work'. Reflecting on this, it is clear that we do not have one single subject that needs to be addressed, but at least four: two individuals; a place; a culture; and a complex of objects. Or, to put things a little differently, we need to:

1. think about the ways in which it is possible to view Julia and Paul as cosmopolitans;
2. examine the degree to which it makes sense to describe the places that Julia and Paul visit as being cosmopolitan in character;
3. explore the ways in which the public culture that Julia and Paul are a part of may be thought of as cosmopolitan;
4. and, finally, we need also to consider how the complex of objects through which places gain a feeling of cosmopolitanism manage to do so.

Which raises the obvious question, where to begin if we are to address these four areas?[1]

> We . . . are a racially and culturally homogenous group of people who have nurtured in isolation from the rest of the world a Victorian, lower-class, Calvinistic, village mentality, and brought it right through the 1970s.
>
> (McLauchlan 1976: 1)

> Other countries have leisure and entertainment industries. Bowling alleys, drive-in brothels, theatres, clubs and other institutions cater for every taste from black current cordial to geisha . . . In New Zealand it's different. Show biz hardly exists outside of a few itinerant pop groups, the Rev. Bob Lowe and that popular group, Dr Geering and the Presbyterian General Assembly . . . As for night life, cities, while beautifully planned, are so laid out you wonder how long they've been dead. A search for the liveliest place in town usually ends up at the YMCA.
>
> (Mitchell 1972: 108)

> [The New Zealand rugby spectator] usually has a good knowledge of geography. He knows where Wales is. He knows Johannesburg is in the Transvaal and that Buenos Aires is the capital of Argentina. He knows the French eat snails and drink a lot of red wine which is alcohol but somehow not real booze. Not men's piss.
>
> (Graham Mourie in King 1988: 117)

Perhaps the most obvious place to start would be the contemporary explosion of interest in the idea of cosmopolitanism in the humanities and the social sciences (see Brennan 1997; Cheah and Robbins 1998; Hutchings and Dannreuther 1999; Tomlinson 1999). Much of this writing is engaging and insightful. Yet there is little work that looks at how cosmopolitanism is actually experienced and incorporated into people's everyday reality. So, rather than working systematically through this extensive contemporary literature on cosmopolitanism, I want to outline a basic framework through which we might begin to think through the everyday presence of the cosmopolitan. This is a framework that draws as much on the classic writings of Georg Simmel, Walter Benjamin, Louis Wirth and Robert Park, and their metropolitan-infected take on the cosmopolitan, as it does on more recent thinking about cosmopolitanism. Nonetheless, I want to begin by following the lead of Ulrich Beck in his book *Der kosmopolitische Blick* (*The Cosmopolitan View*, 2004a; see also Beck 2000, 2002, 2004b), by suggesting the usefulness of drawing a distinction between cosmopolitanism and cosmopolitanization.

Cosmopolitanism

So, let us start with cosmopolitanism. The term 'cosmopolitanism' has a long and varied history in the social sciences dating back to at least the mid-nineteenth century – much longer if we include moral philosophy in the genealogy of these sciences. However, if our focus, as it is in this chapter, is on the lived experience of the cosmopolitan we can isolate two key definitions of cosmopolitanism.

The first definition centres on the idea of cosmopolitanism as a profoundly spatial relationship. Within this definition the cosmopolitan is someone that is at home in the wider world: that is to say someone who moves easily between different cultures and by implication different places. This is the spatial and cultural promiscuity that Ulf Hannerz (1990) described in his famous and much quoted definition of the cosmopolitan attitude in an essay for the journal *Theory, Culture and Society* (and reproduced in his book *Transnational Connections* 1996). For Hannerz the cosmopolitan is defined as someone who has the competence, the patience and the tolerance to live within and understand other cultures. They seek to fit in and make themselves at home wherever they find themselves. The cosmopolitan is someone who is travelled and who wears the knowledge gained from her or his travels lightly. But unlike the locals amongst whom they find themselves, they are always aware that there are other worlds which they could also, and will in all probability at some later time, be a part of. To quote Hannerz (1990: 239):

> Cosmopolitanism . . . is an orientation, a willingness to engage with the other. It entails an intellectual and aesthetic openness towards divergent cultural experiences, a search for contrasts rather than uniformity. To become acquainted with more cultures is to turn into an *aficionado*, to view them as artworks. At the same time, however, cosmopolitanism can be a matter of competence, and competence of both a generalized and a more specialized kind. There is the aspect of a state of readiness, a personal ability to make one's way into other cultures, through listening, looking, intuiting, and reflecting.

It is in this definition of the cosmopolitan, with its focus on the economy of ethnic difference and its obvious connections to processes of globalization, that most of the recent upsurge in interest in the idea of cosmopolitan is rooted. The second definition of cosmopolitanism is more sociological, indeed populist, and has no explicitly spatial dimension (although as we will see it does involve an implicit one). Here cosmopolitanism is also defined as people who are 'of the world'. However, in this second definition this 'of the world-ness' refers to an awareness of the diversity and variety of the immediate world the individual inhabits, coupled to an openness and delight in this immediate variety. Thus, this second definition of cosmopolitanism is very close to being a synonym for urbane-ness, as in someone who is cultivated, refined and at ease in the world, and is intimately linked to the experience of the large city.

One dimension of the variety that the second definition of cosmopolitanism refers to is obviously that of in situ ethnic difference. Equally important is the kind of individual and group variety that is so manifestly part of the metropolitan experience, the diversity that is a product of the sheer density and size of big cities. Large cities are enormous engines for generating difference and variety. As Georg Simmel pointed out over a century ago, perhaps the defining feature of the modern (by which he meant the industrial-bureaucratic) city is the intensity of the division of labour on which it is built and sustained.

> Cities are, first of all, seats of the highest economic division of labour. They produce thereby such extremes as in Paris the remunerative occupation of the *quatorzième*. They are persons who identify themselves by signs on their residences and who are ready at the dinner hour in correct attire, so that they can be quickly called upon if a dinner party should consist of thirteen persons. In the measure of its expansion, the city offers more and more the decisive conditions of the division of labour. It offers a circle which through its size can absorb a highly diverse variety of services. At the same time, the concentration of individuals and their struggle for customers compel the individual to specialise in a function from which he cannot be readily be displaced by another. It is decisive that city life has transformed the struggle with nature for livelihood into an inter-human struggle for gain, which here is not granted by nature but by other men.
>
> (Simmel 1950 [1901]: 420)

The division of labour is not, however, the only source of the individual variety of the metropolis. Parallel to the city's intricate division of labour and the demands that makes on individual distinctiveness, urbanization in a very positive sense also offers the individual the freedom to become utterly themselves. Unencumbered by the tight ties of family and tradition that are in so many ways alien to city life, the city in its density and variety opens up to the individual the chance for all sorts of post-traditional connections. Precisely through its size and the fineness of the differentiation it allows at the social level, the city offers the chance for individuals of any and every particular desire and tendency to find their way to each other. The city creates the possibilities for the exploration and flowering of all sorts of exotic groupings, obscure obsessions, fixations and interests. What is more, the density of urban life and the anonymity that density produces also means that the individual does not have to remain fixed by this identity. As Louis Wirth wrote in his classic essay 'Urbanism as a way of life', in a big city 'no single group has the undivided allegiance of the individual' (2000 [1938]: 101). The urban cosmopolitan self is thus both hybrid *and* fragmented. It is hybrid in that it is made up of a fusion of different identities. And it is fragmented in as much as the successful urbanite is skilled at taking on and taking off a whole series of different temporally bounded identities as they move about the city.

Cosmopolitanization

So that is cosmopolitanism. But if we think back to Julia and Paul's movement through Auckland it is clear that neither of the two definitions of cosmopolitanism outlined above *by themselves* provide much analytical leverage. With both definitions it is easy to gain the impression that either one is open to the world (cosmopolitan) or not (non-cosmopolitan). Hannerz's definition rests (or at least appears to rest) on personal prerogative. The cosmopolitan ethos he is talking about is one that an individual chooses either to embrace or not to embrace. Hannerz's definition is not directly concerned with the question of how we have got to live in a world that appears to be more cosmopolitan. Nor is he very much interested in what that increased cosmopolitan presence might do to the way social bonds and imaginaries are configured – although he is at pains to point out that just because certain social practices appear cosmopolitan that does not mean that they actually are so. The second definition of cosmopolitanism is founded on a much stronger notion that the presence of difference is actually positively generative of a cosmopolitan self (cf. Sennett 1990). It is at least based around the recognition that modern bureaucratic societies generate a kind of internal cosmopolitanism. Nonetheless, if we want to understand the making of this cosmopolitan reality we need a stronger sense of the ways in which cosmopolitan mixings are played out. That is to say, we need to think past cosmopolitanism as a state, towards the processes through which cosmopolitanism comes into existence. We need to think not only about cosmopolitanism; we need also to think about what Ulrich Beck refers to as cosmopolitanization.

Beck defines cosmopolitanization as the process whereby ever more aspects of individuals' and organizations' everyday lives are defined by their connection with things that are not local to it. The local is no longer primarily defined by intimate and well-defined relations. Instead, what we encounter in a particular locality is ever more heterogeneous and pluralistic, and ever less knowable through pure recourse to the local. In much of his writing on cosmopolitanization Beck seems to imply that spatial promiscuity is at the dead centre of the process he is describing. But there is more to Beck's conception of cosmopolitanization than just the effects of greater spatial promiscuity and interconnectedness. Cosmopolitanization is not a straightforward synonym for globalization. Globalization implies a certain grid of homogeneity that allows the world to become global. Indeed, it might be useful to refine Beck's definition above and think of globalization as a particular dimension of the greater interconnectedness of the world that seeks to take heterogeneity and make it manageable and translatable across boundaries (cf. Appadurai's (1996) techno-scapes). That is to say, processes of globalization literally seek to make the world global. In contrast, the concept of cosmopolitanization implies an internal reorganization of social life engendered through the reality of greater diversity.[2] So, while one axis of this diversity is clearly the fact of greater global interconnectedness, another axis is the continued internal (I am tempted to say local) differentiation of social forms (from companies to cities to social groups to nation-states) that is taking place. It is about the ever more complex division of labour that Simmel highlighted over a century ago, and the possibilities for all sorts of creative recombination and reinvention that de-traditionalization brings with it.

But there is also a third axis of diversity that Beck's definition of cosmopolitanization carries within it. This is the diversity and heterogeneity of the material world. This is an aspect that the preceding analytic discussion has barely touched upon. It is not just ideas and people that carry within them the kernel of difference. The material world does so too. Indeed, the material world literally swarms with a heterogeneity that, looked at from a certain angle, makes every social relationship cosmopolitan (see De Landa 1997; Clark 2000). Nonetheless, I would like to stress two key elements of the material world's cosmopolitan force. First, the material can itself be straightforwardly cosmopolitan in the sense that it invokes, or speaks through, its very substance of a certain worldliness. A whole range of materials and assemblages of materials are defined by careers of movement; they bear within them (either explicitly or implicitly) a whole geo-history of movement, displacement and reinvention. The most obvious examples are of course all sorts of plants and foodstuffs –

tomatoes, sugar, coffee, tea, pepper – that have profoundly reshaped not only what the world eats, but how the world *tastes*. Second, the material can also be viewed as cosmopolitan in the way that it demonstrates the brute alterity, or otherness, of the world.

So, to summarize and to slightly reinterpret and reconfigure the argument that Beck develops in *Der kosmopolitische Blick*, to think about cosmopolitanization is (1) to think about the ways in which the world we live in is organized through all sorts of 'machines' for generating heterogeneity (cf. Latham and McCormack 2004; Amin and Thrift 2002). And (2) in light of this emergent heterogeneity, it is to think about the ways that this heterogeneity is incorporated (or indeed repudiated) within existing social forms. To think about cosmopolitanization is to start thinking about the ways that the world actually *becomes* cosmopolitan.

So let us try and think about the particular ways that the world has become cosmopolitan in parts of Auckland, and how this cosmopolitan-ness has become part of many people's everyday routines. To do this, along with the already introduced Paul Rennie Brown I want to focus on the lives of the two women Julia and Katja – two divorced 'solo' mothers in their early 40s – whose Wednesday routine is sketched out in Figure 5.3. Neither Julia or Katja are obvious cosmopolitans. Julia lives in a rented Herne Bay flat with her two school-age children. Her working day involves organizing the care of her children, the production of her handmade prints, and working at finding new places to sell what she has made. Katja has three children, all of whom are also of school age. Like Julia she organizes a range of jobs around the task of raising her children. And also like Julia, after many years of home ownership, she is currently living in rented accommodation. At one level the rhythm of both their lives appears to resemble those of classic suburban parents: lives organized around the routines of childcare, work, shopping and socializing with similarly aged friends. Both are good friends with Paul Rennie Brown, with whom they sometimes share both a bottle of wine and baby-sitting. Yet, if we look carefully not only at their movement around Auckland on the Wednesday sketched out in the figure, but also more generally at the rhythms of how they inhabit Auckland, Julia and Katja's lives reveal a range of remarkably cosmopolitan attributes. The problem is finding the right key in which to narrate this cosmopolitanism.

Let us begin with what we might call Julia and Katja's 'life in public'. A significant proportion of both Julia's and Katja's movement around

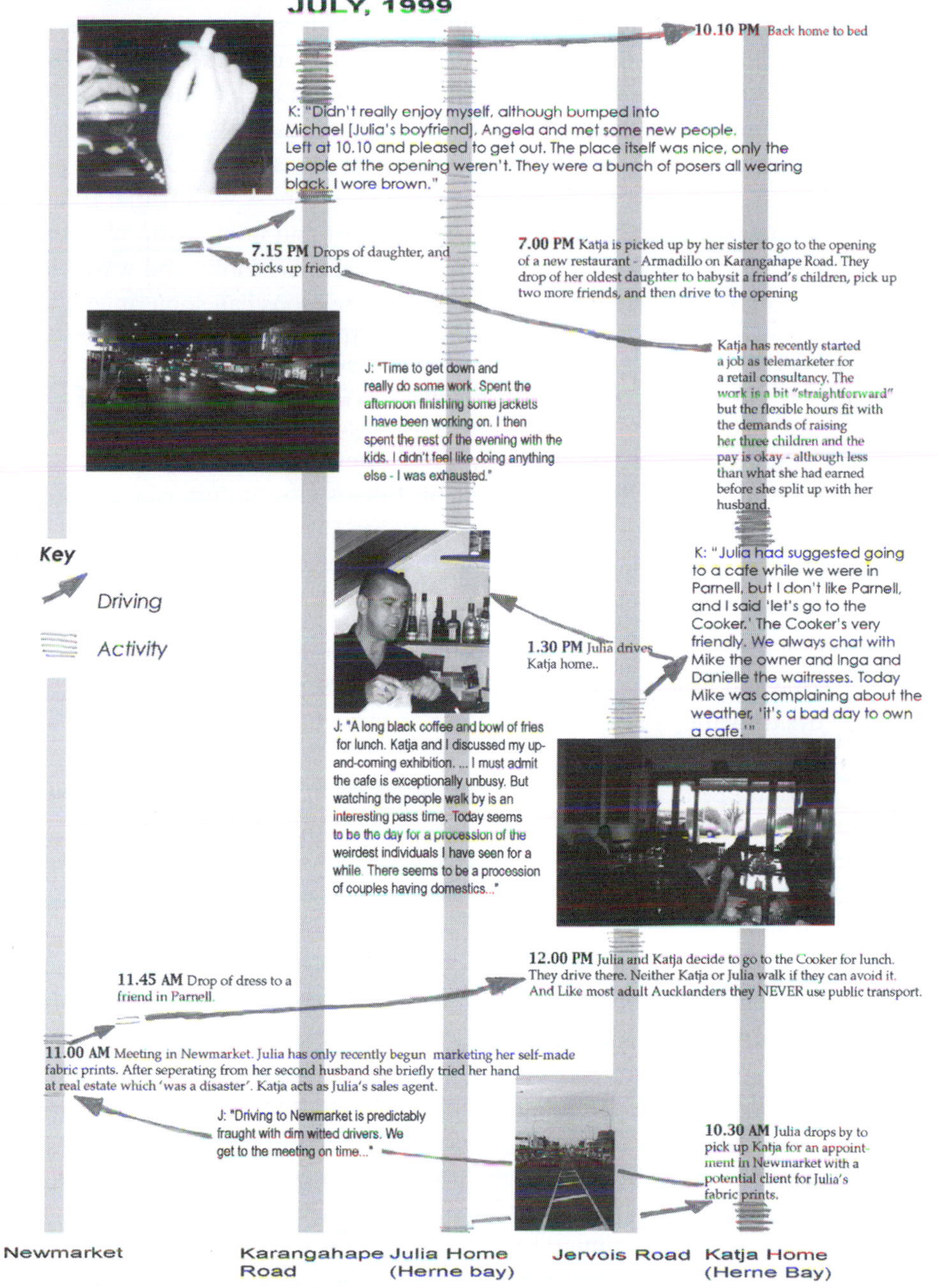

Figure 5.3 ***An ordinary Wednesday, July 1999.***

Auckland is a consequence of either childcare or work demands. Nonetheless, around those commitments Julia at least also has a great deal of freedom about how she organizes her time. On days when she is working at home, she also likes to overcome her isolation by meeting up with friends for lunch and cigarettes at the Cooker, a smart if a little austere licensed café on nearby Jervois Road. Once there, she may use the opportunity to meet and talk business with Katja. But just as often she and whoever she is with just sit and talk, gossip with Inga the Cooker's waitress and Michael its owner, or smoke and watch the world go by. If the mood takes her these sessions can sometimes stretch out to a couple of hours and a bottle of wine.

The Cooker is by no means the only place that Julia, or Katja, or their good friend Beth use. They also occasionally go to the other cafés next to the Cooker on Jervois Road. They often visit nearby Ponsonby Road for a drink or a evening meal. Julia is also enthusiastic about Armadillo. What is perhaps most striking about Julia's use of the Cooker and similar kinds of places is simply the ease with which she inhabits them – and indeed shares them with other men and women.

Writing about the nineteenth-century illustrator Constant Guy, the critic and lyric poet Charles Baudelaire (1992: 399–400) wrote of the joys of being at home in the public spaces of the city:

> To the perfect idler, the passionate observer, it becomes an immense source of enjoyment to establish his dwelling in the throng, in the ebb and flow, the bustle, the fleeting, and the infinite. To be away from home, and yet to feel at home.

As Baudelaire (1992: 395) argues, such embodiments of flâneuristic idleness are perfect 'men of the world'. By which he means not only that such figures are knowledgeable about the world around them. They are also insatiably curious about it. Now, clearly Julia and her occupation of the public spaces of Auckland does not fit exactly with Baudelaire's description of the cosmopolitan idler. In Baudelaire's description the idler is a profoundly solitary and disconnected figure. He (and it is always a he in Baudelaire) is someone who watches and observes without aiming to connect (or only to connect in the imagination). Julia and Katja's relationship to the activities of the street are much more socially connected. Julia knows the local 'public characters' (Jacobs 1961: 394), and she says that she would not generally go to a café alone. She likes to watch the world go by *with* other people. But like the idler she and her friends actively, and confidently, participate in what we might call, following Lynda Nead (2000: 71), the city's 'ocular economy'. That is to say, an interplay of glances and gazes, of looks and eye contact from which the participants take a certain pleasure – and which often spills over into gossip between those doing the watching. As Julia says, all the different kinds of people, ways of walking and dressing, all the different kinds of bodies that are on display even on the relatively quiet Jervois Road pavement are fantastically stimulating. It is a 'wonderful stretch to watch people'. If nothing else it is an activity that makes life *interesting*.

Now we could at this point consider in more detail the structure of this ocular economy of which Julia and Katja are a part. We could talk about how it is organized through a certain valorization of novelty and diversity. Readers will recall that Hannerz's (1990: 239) description of the cosmopolitan attitude rested on 'an intellectual and aesthetic openness'. And, indeed, one of the most interesting elements of this particular ocular economy is precisely that at least aesthetically it is organized around a profoundly *cosmopolitan* view. It is a way of looking that carries the very *urban* gaze of the flâneur into the suburbs of Auckland, even if only to the inner suburbs. But perhaps what is the most interesting element of this ocular economy is the shared – if implicit – recognition of how it works. Talking about his use of places like those along Jervois Road, Julia and Katja's friend Paul Rennie Brown tells a similar – if more obviously sexualized – account of the pleasures of looking and being looked at by others. But he is also clear that this is a pleasure that is built around a respect for the unwritten rules that allow spaces like the Cooker to operate in the way they do. It involves tactfully respecting the space of others, and judging when people can be approached or not. Contrary to Hannerz's suggestion that the cosmopolitan is dependent on the presence of locals, it is precisely the presence of a certain pluralistic *ambiance* that structures the kind of cosmopolitanism we find along Jervois Road and elsewhere in inner-city Auckland. To make sense of how this structuring works we need to leave Julia, Katja and Paul for a while and consider just how it is that the complex of objects – the coffee, the bottles of wine, the sidewalk tables, the bowls of *French* fries and so forth – assembled along places like Jervois Road came to have the feel, the ambiance that they do. And to do this we have to move back in time to a different New Zealand.

> I wandered along looking at the girls and wondering what the hurry was. I nearly got run over a couple of times so I took refuge in the pub. A very friendly bloke called Joe said he could see I was a stranger and came over to talk to me. I appreciated the thought even though it cost me his drinks and the drinks of his relatives who seemed innumerable.
>
> That night I ended up at a house where about £20 worth of my beer was being noisily drunk by several dozen people I had never seen before. Not that I cared much – I was lying ill in a flower-bed, calling loudly for the dog. Eventually somebody carted me inside and put me to bed. I swore never to touch the stuff again, and solemnly advised the women who tucked me in to follow my example.
>
> (Crump 1960: 162)

> Give us men,
> Strong and stalwart ones,
> Men who highest hope inspires,
> Men who purest honour fires,
> Men who trample self beneath them,
> Men who make their country wreath them,
> As her noble sons
> Worthy of their sires;
> Give us men.
>
> (Salvation Army song, late nineteenth century, in Phillips 1987: 63)

> [Drinkers are] packed into the central area [of the bar] at the rate of about one person per square foot, and standing shoulder to shoulder, elbowing past one another, reaching over one another's heads, and spilling brimful glasses over one another's clothes, they absorb glass after glass of amber fluid. Their natural thirst is intensified by the vertical position (because there are patently not enough seats to go around, and sitting down tends to handicap the drinker in the race for a refill), and also by the anxiety induced by the certain knowledge that the moment the clock strikes six, a harsh jangling of bells will signal the abrupt cessation of the flow of beer.
>
> (Bollinger 1967: 3–4)

To think back to New Zealand society in the 1960s and early 1970s is in many ways to cross a historical and cultural divide. Social commentators and historians writing about New Zealand society in the immediate post-World War II decades describe a society defined by its homogeneity. It was a society where, as the historian Bill Oliver wrote (1960: 275), even in the biggest cities Auckland and Wellington 'the *ethos* of the small town, where back fence and main street gossip executes the minority before it can justify itself, still dominates'. It was also a society marked by a strongly reformist zeal. While the primary motivation of most European settlers to New Zealand had been a concern with brute material welfare – they wanted to be materially better off than they had been – this was twinned with a faith in the *newness* of the country they were setting sail for. A new society could, and for many *should*, be a much better society than that which had been left behind.

This deeply pragmatic and utilitarian reforming zeal has left a lasting – and surprisingly distinctive – imprint on New Zealand society in general, and on the nation's urban public culture in particular. It is impossible in the space of this chapter to fully summarize the scope and distinctiveness of this social programme.[3] However, following the lead of the social historian Miles Fairburn (1975: 6), who argued that nineteenth-century New Zealand settler society was driven by the urge to create a kind of 'familial Arcadia', we might usefully describe New Zealand as having been defined by a very peculiar Arcadian modernism.

This Arcadian modernism was deeply suspicious of both the urban and the cosmopolitan.[4] Worried about the – apparently – socially corrosive nature of urban life, particularly those public traditions which centred around the consumption of alcohol, New Zealand society set about elaborating a baffling complex set of rules to control the evolution of much of its adult urban public culture. It was these rules that created the sense of deadness captured by Austin Mitchell in his 1972 caricature of New Zealand society, *The Half-Gallon, Quarter-Acre, Pavlova Paradise*, where, not coincidentally, 'the search for the liveliest place in town ends up at the YMCA' (1972: 108). Ironically it was also these rules that had allowed a brutal, intensely masculine, colonial drinking culture that centred on the hotel bar and rugby club to survive with remarkable vigour into the middle of the twentieth century. This was a culture where beer is not beer, it is 'piss', where women are not women, they are 'sheilas', 'lays' and 'nags' (Phillips 1987; King 1988). What is more, at the same time these rules effectively closed down many opportunities for the development of a less aggressively masculine urban public counter-culture (or indeed a feminine one).

Now it would not be accurate to say that New Zealand cities were as entirely monocultural as Oliver's or Mitchell's accounts suggest. It is possible to map out an alternative history that would highlight a range of diverse groups and organizations that did not fit into the Arcadian suburban ideal. But the point is that almost all of these – with some exceptions – were hidden from obvious public view. What is also of central importance to our story is that when in the late 1960s and 1970s a range of groups, from women to intellectuals, homosexual men and the professional urban middle classes, began to campaign for the right to inhabit the city in ways different to the then prevalent norms, they did not encounter a landscape rich with opportunities for sociality. It is within this context that the kind of cosmopolitanization we have encountered in places like Armadillo, the Cooker and other similar places becomes significant.

In seeking to articulate the possibilities of what a more open and less tightly regulated public urban landscape might be like, activists seeking reform drew heavily on models from outside the country. In part this involved little more than pointing out that many of New Zealand's regulations – particularly those controlling the public selling and consuming of alcohol – were more restrictive than almost anywhere else in the world (Bollinger 1967; Belich 2001). But it also involved mobilizing a range of narratives about the urban public cultures of a number of continental European societies. These narratives were used to suggest that the form and ambience of New Zealand cities were neither desirable nor a necessary product of the New Zealand 'character'. What is most interesting about these narratives is that while they were 'cosmopolitan', in as much as that they involved drawing on ideas from outside New Zealand, they simultaneously involved an attempt to articulate an alternative narrative about what New Zealand culture was about. While they drew on a highly stylized and often idealized view of places like France and Italy, they did so not because of any desire to become French or Italian but to extend the range of possibilities for being a New Zealander.

But, perhaps more importantly, when a series of radical regulatory reforms in the late 1980s finally made it possible to open a café or a restaurant or a bar without a range of expensive institutional hurdles, entrepreneurs found not only a significant latent demand. They also found themselves working in an environment in which they had few local examples to draw upon. In an attempt to plug into those markets not being catered for by the traditional hotels and pubs – or, indeed, the existing coffee shops and tea houses – many entrepreneurs drew with varying degrees of fidelity on blueprints from outside the country. In essence these entrepreneurial ventures represented a series of on-going grassroots experiments to reinvent the country's urban public culture.

Figure 5.4 ***Glasses.***

Within the context of this experimentation the presence of a whole range of objects that had previously been foreign to existing hospitality spaces served two purposes. First, they offered a kind of counter-narrative to existing hospitality spaces. The presence of everything from French-style bistro tables and expensive Italian espresso machines to chalkboard menus,[5] lattes and spiralina drinks set down a series of physical and semiotic markers that these spaces were meant to be used in quite different ways from existing spaces. While the presence of these objects did not exclude the use of these new hospitality spaces by regular pub or hotel users (that is to say, men), they did quite explicitly demand that they must use them in a quite different way.

Second, they facilitated a reconfiguration of the relationship between a whole range of materials and the social norms that shaped their use. Perhaps the most striking example of this was the relationship of alcoholic consumption and public space. The presence of alcohol for sale ceased to be *the* singular defining feature of a hospitality space. The idea that people might go to a place that had a liquor licence and actually drink something other than alcohol involved a complete reordering of the idea of what alcoholic drinks were for, and indeed how one should behave in hospitality spaces where alcohol was served. But – equally importantly – the incorporation of simple architectural features like bi-folding and French windows into the design of many cafés and bar-restaurants along with street seating also completely

Figure 5.5 ***Jervois Road.***

changed the relationship of these new hospitality spaces to the street. Where previously hospitality spaces were oriented inwards and screened from the street outside, the dominant orientation of many of the new spaces is towards the street, playing up to and offering a seat within the wider public theatre of the street.

Looking back at the transformation narrated in the previous section it is possible to see the emergence of the kind of social spaces described in this chapter as being nothing but a positive good. That certainly is the popular interpretation (see Burton 1994). However, it has not been the aim of the preceding discussion to suggest that the new urban culture is straightforwardly superior to the one that preceded it. There are qualities of the older public culture which are worth admiring – its egalitarianism (between men at least), its lack of pretension, its rooting in a set of easily identifiable collective values (Sinclair 1961; Oliver 1981; Belich 2001). The cosmopolitan is not inherently better than the parochial. Rather, what is interesting about the cosmopolitan is how it is put to work, how it is employed in the making and remaking of particular social practices and particular places.

And this brings the chapter back to the question with which it started. How should we understand the cosmopolitanization of certain parts of

Figure 5.6 ***Tuatara.***

inner-suburban Auckland? Well, if we start with Julia, Paul and Katja, it is not at all clear to what degree it makes sense to talk of them as cosmopolitans. They did not describe themselves as cosmopolites in their diaries and interviews. When they talked about their lives, and how they organize them, they generally described an intensely parochial horizon of action. While Julia and Katja are well travelled – if Paul is not – in each case their focus is very much on what is going on in Auckland. And it is from within this perspective that they interpret and make use of the kinds of spaces being opened up to them along places like Jervois and Ponsonby Roads. Theirs is an implicit cosmopolitanization. It is their presence, and style of inhabitation, that is significant.

What has very clearly and unambiguously become cosmopolitan is the material culture of spaces like Jervois and Ponsonby Roads. Here the point is not that the global, something inherently alien and foreign, is asserting itself onto the local landscape. Instead what we can see – in the case examined in this chapter at least – is how certain select materials and ideas from elsewhere are used to reconfigure in various ways how particular local materials, particular local relationships, particular kinds of local spaces, can be understood. Indeed, what is perhaps most interesting about the new urban public culture that has emerged in Auckland is how much the 'cosmopolitan' is in fact rooted within the boundaries of the national. Many of the key materials – the wine, the ingredients for the food, the buildings, the streetscape, to name a few examples – of this public culture were present prior to the 1990s. The process of cosmopolitanization involved a reinterpreting and revalorizing of a series of previously neglected and undervalued local materials and cultural practices. What is more, somewhat ironically, the process of cosmopolitanization also opened up a way of re-reading New Zealand's contemporary history that recognizes that historically New Zealand society was in many ways more diverse, and more complex, than earlier national imaginaries had acknowledged.

And this – perhaps – is the greatest strength of Beck's concept of cosmopolitanization. It allows us to focus upon how particular connections across often great distances come to have certain resonances, and certain affective qualities, in particular places. More importantly, it manages to bring the far away into focus in such a way as to avoid setting up a series of a priori hierarchies between the 'local', the 'national' and the 'global'. If we want to understand how the cosmopolitan and the global are abroad in our worlds it is just these kinds of grounded stories suggested by Beck's approach and explored in this chapter that we need to develop.

Acknowledgements

I would like to thank all those people who gave their time to aid my research – especially the diary writers whose interaction with Auckland is recounted here. The research for the project was funded by the New Zealand Foundation for Research, Science and Technology. I would like to thank NZFoRST for their generous funding. I would also like to thank Professor Richard Le Heron for his support during the project, and Jon Binnie, Julian Holloway, Steve Millington and Craig Young for offering me the opportunity to write this chapter. Lastly, I would especially like to thank Shaun French for his suggestions on cutting the chapter down to a suitable length for publication. The usual disclaimers apply.

Notes

1 The research discussed in this and the following two sections summarizes part of a project called 'Auckland's Changing Urbanities' undertaken by the author

between October 1997 and October 1999. The research was funded by a post-doctoral fellowship from the New Zealand Foundation for Research, Science and Technology. The discussion of Paul, Julia and other café/restaurant/bar users in this chapter is based on ethnographic research on regular users of Ponsonby and Jervois Roads' hospitality establishments. In total 38 people wrote one-week personal diaries, took photos and gave in-depth interviews based on their weekly diaries. In addition, another 12 people wrote diaries for another site. Many of these respondents also used Jervois and Ponsonby Roads. For an in-depth discussion of the methodology used, see Latham (2003a, 2004). Additional information is based on interviews with 35 owners or co-owners of restaurants, cafés or bars, in combination with the analysis of site files from the Auckland District Liquor Licensing Authority, Auckland City Council property site files, and newspaper and magazine articles. The story presented is also informed by a number of interviews with Auckland City Council planners, local politicians, community activists and long-term residents (see Latham 2000, 2002, 2003b).

2 To put things very simply, the 'cosmopolitan' is the mix of diversity while the 'global' is the instrument of translation/consistency that allows this mixing to take place across time-space.
3 See Sinclair (1961), Oliver (1981) and Belich (2001). Thorough though all these accounts are, and although written by authors resident in New Zealand, none really catches the weirdness *and* distinctiveness of the New Zealand society that emerged at the end of the nineteenth century and the start of the twentieth.
4 It is of course deeply paradoxical that a society that was from its foundation highly urbanized should be built on such strongly anti-urban sentiments (Franklin 1978). It is also paradoxical that a society that was dependent in so many ways – for its prosperity, for its population, for many of the material goods that defined the Kiwi good life – on its traffic with the rest of the world should place so much emphasis on restricting and controlling this traffic. A sense of how deeply rooted this Arcadianism was can be seen in the fact that when New Zealand began its first wave of mass public housing construction in the 1930s almost all of this housing consisted of detached bungalows set on quarter-acre lots.
5 Chalkboards were in fact forbidden by public health regulations until the mid-1980s because it was feared the chalk dust could contaminate food. For a similar reason sidewalk tables were also banned. The fact that the rest of the world was not succumbing to food poisoning en masse despite the widespread prevalence of both these practices seems to have passed New Zealand's public health authorities by.

References

Amin, A. and Thrift, N. (2002) *Cities: Reimagining the Urban*, Cambridge: Polity Press.

Appadurai, A. (1996) *Modernity at Large: Cultural Dimensions of Globalization*, Minneapolis, MN: University of Minnesota Press.

Baudelaire, C. (1992) *Baudelaire: Selected Writings on Art and Literature*, London: Penguin.

Beck, U. (2000) 'The cosmopolitan perspective', *British Journal of Sociology*, 51 (1): 79–105.

Beck, U. (2002) 'The cosmopolitan society and its enemies', *Theory, Culture and Society*, 19: 17–44.

Beck, U. (2004a) *Der kosmopolitische Blick oder: Krieg ist Frieden*, Frankfurt am Main: Suhrkamp Verlag.

Beck, U. (2004b) 'Mobility and the cosmopolitan society', in W. Bonß, S. Kesselring and G. Vogl (eds), *Mobility and the Cosmopolitan Perspective*, Munich: Universität der Bunderswehr München.

Belich, J. (2001) *Paradise Reforged: A History of the New Zealanders from the 1880s to the Year 2000*. Wellington: Penguin.

Bollinger, C. (1967) *Grog's Own Country: The Story of Liquor Licensing in New Zealand*, Auckland: Minerva.

Brennan, T. (1997) *At Home in the World: Cosmopolitanism Now*, Cambridge, MA: Harvard University Press.

Burton, D. (1994) 'Introduction', in *Character Cafés in New Zealand*, Wellington: Phantom Books.

Cheah, P. and Robbins, B. (eds) (1998) *Cosmopolitics: Thinking and Feeling Beyond the Nation*, Minneapolis, MN: University of Minneapolis Press.

Clark, N. (2000) '"Botanizing the asphalt?" The complex life of cosmopolitan bodies', *Body and Society*, 6: 12–33.

Crump, B. (1960) *A Good Keen Man*, Wellington: Reed.

De Landa, M. (1997) *A Thousand Years of Nonlinear History*, New York: Swerve Editions.

Fairburn, M. (1975) 'The rural myth and the new urban frontier: an approach to New Zealand social history, 1870–1940', *New Zealand Journal of History*, 9: 3–21.

Franklin, H. (1978) *Trade, Growth, and Anxiety: New Zealand Beyond the Welfare State*, Wellington: Methuen.

Hannerz, U. (1990) 'Cosmopolitans and locals in world culture', in M. Featherstone (ed.), *Global Culture*, London: Sage.

Hannerz, U. (1996) *Transnational Connections: Culture, People, Places*, London: Routledge.

Hutchings, K. and Dannreuther, R. (eds) (1999) *Cosmopolitan Citizenship*, Basingstoke: Macmillan.

Jacobs, J. (1961) *The Death and Life of Great American Cities*, New York: Norton.

King, M. (ed.) (1988) *One of the Boys? Changing Views of Masculinity in New Zealand*, Auckland: Heinemann.

Latham, A. (2000) 'Urban renewal, heritage planning and the remaking of an inner-city suburb: a case study of heritage planning in Auckland, New Zealand', *Planning Practice and Research*, 15: 285–98.

Latham, A. (2002) 'Re-theorizing the scale of globalization: topologies, actor-networks, and cosmopolitanism', in A. Herod and M. Wright (eds), *Geographies of Power: Placing Scale*, Oxford: Blackwell.

Latham, A. (2003a) 'Research, performance, and doing human geography: some reflections on the diary-photo diary-interview method', *Environment and Planning A*, 35: 1993–2017.

Latham, A. (2003b) 'Urbanity, lifestyle and making sense of the new urban cultural economy', *Urban Studies*, 40: 1699–724.

Latham, A. (2004) 'Researching and writing everyday accounts of the city: an introduction to the diary-photo diary interview method', in C. Knowles and P. Sweetman (eds), *Picturing the Social Landscape*, London: Routledge.

Latham, A. and McCormack, D. (2004) 'Moving cities: rethinking the materialities of urban geographies', *Progress in Human Geography*, 28: 701–24.
McLauchlan, G. (1976) *The Passionless People*, Auckland: Cassell New Zealand.
Mitchell, A. (1972) *The Half-Gallon, Quarter-Acre, Pavlova Paradise*, Christchurch: Whitcombe and Tombs.
Nead, L. (2000) *Victoria Babylon: People, Streets and Images in Nineteenth-Century London*, New Haven, CT: Yale University Press.
Oliver, W. H. (1960) *The Story of New Zealand*, London: Faber and Faber.
Oliver, W. H. (ed.) (1981) *The Oxford History of New Zealand*, Wellington: Oxford University Press.
Phillips, J. (1987) *A Man's Country? The Image of the Pakeha Male*, Auckland: Penguin.
Sennett, R. (1990) *The Conscience of the Eye*, London: Faber and Faber.
Simmel, G. (1950 [1901]) 'The metropolis and mental life', in K. Wolff (ed.), *The Sociology of Georg Simmel*, New York: Free Press.
Sinclair, K. (1961) *A History of New Zealand*, Auckland: Penguin.
Tomlinson, J. (1999) *Globalization and Culture*, Cambridge: Polity Press.
Wirth, L. (2000 [1938]) 'Urbanism as a way of life', in R. Legates and F. Stout (eds), *The City Reader*, third edition, London: Routledge.

6 Cosmopolitanism by default: public sociability in Montréal

Annick Germain and Martha Radice

Introduction

Cosmopolitanism crops up regularly in debates about modern societies, and over the last few years social scientists have been exploring it afresh. Historically, the concept was used in opposition to nationalism, particularly by the anti-Semitic right in France at the turn of the last century who denounced the superficiality and rootlessness of international elites (Winock 1997). Vertovec and Cohen (2002) identify several trends among the recent perspectives on cosmopolitanism. It has been understood as an individual attribute of openness towards other cultures, as a project uniting political bodies across national borders and as a project of citizenship that can cope with subjects' multiple affiliations, especially their ethnic or cultural ones. Some therefore see cosmopolitanism as an alternative to 'tired' models of multiculturalism. Globalization frames these discussions and brings out the urban dimensions of cosmopolitanism, given the concentration of ethno-cultural diversity in large cities and the role that cities play in advanced economies.

Montréal provides a particularly interesting context for investigating cosmopolitanism, and not only thanks to its ethno-cultural diversity. It is the metropolis of a province, Québec, whose majority ethnic population (white francophone descendants of the settlers of Nouvelle-France) contests its minority status within the predominantly 'Anglo' confederation of Canada. The old opposition between nationalists and cosmopolitans (Cheah 1998) remains pertinent here. While the political agenda is encumbered with *la question nationale*, cosmopolitanism plays a special role as an urban affair in Montréal. In this chapter, we explore the meaning of cosmopolitanism in the everyday life of this city, focusing on its multiethnic dimensions. We visit Mile End, one of seven multiethnic Montréal neighbourhoods studied as part of a large-scale research project on sociability in public spaces. We show how a shared – if 'fuzzy' – representation of cosmopolitanism supports successful interethnic cohabitation in this neighbourhood.

Cosmopolitanism as urban culture

Few authors have related cosmopolitanism explicitly to urban spaces and places, but we identify two camps amongst them. Crudely, there are the cynics who see it as an elitist ideology that strategically co-opts 'cultural' difference in order to sell experiences to urban consumers; and the idealists, who still believe in its Enlightenment-inspired progressive potential to unite citizens and/or political movements and institutions across national and other boundaries. These trends are apparent in three recent articles and a somewhat older book (Kahn 1987; Hiebert 2002; Law 2002; Binnie and Skeggs 2004). Binnie and Skeggs (2004) explore the branding of Manchester's Gay Village as cosmopolitan. They follow Hannerz's definition of cosmopolitanism as a 'willingness to engage with the Other', leading to the development of a cosmopolitan 'competence', 'a state of readiness, a personal ability to make one's way into other cultures, through listening, looking, intuiting and reflecting' (1996: 103). However, this competence is unevenly attributed. The privileged few who acquire it are people who feel at least partially at home everywhere, adapting to the culture in which they find themselves. Thus, straight white middle-class Mancunians learn how to move 'respectfully' through the Gay Village and enjoy its uniqueness and difference, while their discourse excludes certain groups (such as white working-class people) from achieving cosmopolitan competence. Even the gay people who created the space 'can only be cosmopolitan by existing as a sign of difference for others' (Binnie and Skeggs 2004: 53).

Since Binnie and Skeggs (2004) understand cosmopolitanism as an individual attribute, the urban space they study is cosmopolitan only in that it is apprehended through knowing and consuming the 'Other'. Daniel Hiebert's (2002) interpretation of cosmopolitanism in Vancouver is more idealistic. He too begins with Hannerz's (1996) openness to otherness, but sees this as leading to a capacity not to consume but rather 'to interact across cultural lines' in places 'where diversity is accepted and rendered ordinary' (Hiebert 2002: 212), such as his own multiethnic neighbourhood. He opposes cosmopolitanism to transnationalism. The latter connotes experiencing and being attached to two or more places in different nation-states, while the former facilitates meaningful exchange between these places or their cultures. Hiebert and his colleagues found that immigrants and their children in greater Vancouver switch back and forth between mainly transnational and mainly cosmopolitan social settings. The 'progression' from transnational to cosmopolitan, often seen as an indicator of successful integration, is neither obvious nor one-way.

Hiebert's cosmopolitanism still seems to be constituted – like Binnie and Skeggs' (2004) version – in the sphere of individual (inter)action. Lisa Law's article (2002) makes explicit associations between the postcolonial urban landscape of Hong Kong and cosmopolitanism as a political project. She describes how transnational Filipino labourers, specifically women working in long-term domestic service, take over the prime commercial space of Hong Kong's Central district on Sunday, their day off. They gather to socialize, share food, write and read letters and buy cheap goods from home in spite of the (literal) barriers erected by the main landowner, Hong Kong Land, who would rather they kept moving (and spending in designer shops). The district's once a week transformation into 'Little Manila' has made it a site for political mobilization, as non-governmental organizations (NGOs) use the space to campaign for migrant workers' rights. Pro-democracy and environmental movements publicize their concerns there too, and the Sunday gatherings have become meaningful for non-Filipino Hong Kong residents (Law 2002: 1639–42, 1636–7). Thus, they transcend the transnational purpose of meeting with compatriots. They imbue the Central district's landscape with new senses of place and constitute a political, public sphere that, crucially, 'cannot be reduced to Hong Kong or the Philippines' (Law 2002: 1641–2). This is cosmopolitanism as a collective political project, in both the senses that Vertovec and Cohen (2002) sketch out. It is part of a 'cosmopolitanism from below', an 'emerging global civil society' of transnational social movements, and it also mediates some of the multiple subject affiliations of the people involved (as transnational workers, service workers, Filipinos, Hong Kong dwellers, mothers, daughters and so on).

Finally, we turn to a sociological essay with a humanist flavour published in 1987, in which Bonnie Menes Kahn draws on the history of world cities like Babylon, Constantinople, Vienna and New York to identify conditions that a city should fulfil to be considered cosmopolitan. It should have a culture of tolerance built on a richness of public life, particularly in animated public spaces that belong to everyone, where the city's variety is obvious and observable (Kahn 1987). A cosmopolitan city's ethno-cultural diversity should be reflected in the opportunities it gives 'strangers' to succeed economically and to participate fully in city life. Lastly, a cosmopolitan city should have 'a vision, a sense of purpose or mission' (Kahn 1987: 17) which, although vague, evokes the political aspects of cosmopolitanism discussed above. These criteria, along with the themes of the three articles, provide a useful heuristic for thinking about what cosmopolitanism means in the context of Montréal.

Montréal as a cosmopolitan city

Montréal's great range of ethno-cultural diversity is indisputable today, but it bloomed later than in the big American cities. Having been a trading post for indigenous peoples in pre-contact times (pre-1500), Montréal remained a small French missionary village, Ville-Marie, during the first century of its colonial existence from 1642. It became a city of multiple cultures only after the British conquest of New France (1759), when English, Irish and Scottish immigrants settled near the original French inhabitants. All these groups differed in many ways from each other (language, religion, colonial outlook and allegiances, socio-economic status) and they established themselves in different neighbourhoods as a result. In 1901, people who could trace their origins back solely to France or the British Isles still made up 96 per cent of the island's population (Linteau 2000), but the first sizeable waves of non-French, non-British Isles immigrants soon arrived.

Until the 1970s, they were mainly European (although there were significant Chinese and African-Caribbean communities in Montréal by the early twentieth century), and they tended to settle in distinct residential areas, such as Little Italy and the Greek and Portuguese neighbourhoods (McNicoll 1993). Since the end of the 1970s immigrants' origins have diversified, such that newcomers from Europe form a minority in recent influxes while others hail from parts of Africa, Asia, the Caribbean, Latin America and the Middle East.

The immigrant population (permanent residents who were not born in Canada) is decidedly smaller in the city of Montréal, where it makes up 28 per cent of the general population, than in the cities of Toronto (49 per cent) or Vancouver (46 per cent) (Statistics Canada 2004). But its origins are more diverse, principally because in addition to the groups that immigrate to other Canadian cities, Montréal attracts people from countries where French is a typical second language, including Romania, Vietnam, Haiti and North African countries. There has also been a recent revival of immigration from France for the first time since the eighteenth century (Table 6.1). We shall return to the topic of this *francophonie* later.

Montréal's multiethnicity is now inscribed in the fabric of the mixed neighbourhoods that have come to replace the old 'ethnic villages'. This means that most Montréalers cross paths daily with people from a great diversity of cultures, in both inner-city neighbourhoods and the inner suburbs. Up until the mid-1980s, immigrants integrated easily into the job market, and often attained a higher level of economic success than people born in the country. Over the past decade, this trend has changed. While many immigrants still prosper, others have a great deal of difficulty entering the job market, in spite of – or perhaps because of – their qualifications and credentials (over 65 per cent of immigrants who arrived in the province of Québec between 1999 and 2003 had more than 14 years of formal education (Gouvernement du Québec 2004)). Members of linguistic and visible minorities (defined

Table 6.1 ***Top ten countries of origin of immigrants to the metropolitan region of Montréal during the 1990s***

Country of origin	*% of total number of immigrants*
Haiti	6.6
China	6.4
Algeria	5.8
France	5.8
Lebanon	4.9
Morocco	4.1
Romania	3.7
Philippines	3.5
India	3.4
Sri Lanka	3.3
Total top ten	47.5

Source: Statistics Canada

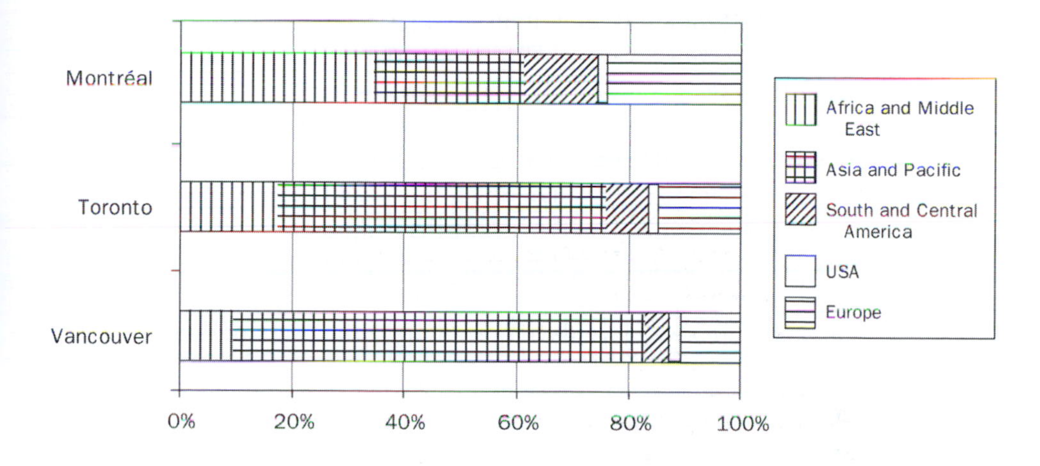

Figure 6.1 ***Immigration by source area to metropolitan regions, 2001.***
Source: Citizenship and Immigration Canada (2002).

under Canada's Employment Equity Act as 'persons, other than aboriginal peoples, who are non-Caucasian in race or non-white in colour' (Department of Justice 2004)) are seriously under-represented in the public sector, despite programmes designed to increase their access to these jobs. In 1999, only 7.4 per cent of the municipal workforce came from a minority background, compared to around 40 per cent in the city's population (Ville de Montréal 2000). In addition, members of visible minorities (whether immigrants or not) are over-represented in the unemployment statistics (Jedwab 2003). This does not go as far as what political scientist Daniel Latouche (1997: 3) calls 'truncated cosmopolitanism' (*cosmopolitisme tronqué*), in which foreigners are confined to unattractive job sectors, but the trend is worrying and jeopardizes any claim that Montréal is 'truly' cosmopolitan, if one accepts Kahn's criterion of opportunity for economic success.

Public space and sociability

In her chapter on Vienna, Kahn (1987) demonstrates the importance of the Prater, a large urban park commissioned by Emperor Joseph II in 1765 and designed to bring together people of all social classes. A similar principle underlies the North American parks designed by Frederic Law Olmsted, which include the Parc Mont-Royal, Montréal's own iconic green space. Montréalers of all stripes use this park on the flanks of the pint-sized 'mountain' that gives the city its name (Debarbieux and Perraton 1998). Forms of public sociability that demonstrate the everyday negotiation of diversity also exist at the local level. Neighbourhood public spaces can be good places to observe how a multiethnic city operates, especially if people frequent them out of choice rather than obligation (Remy 1990). We might surmise along with Kahn (1987: 125) that '[p]erhaps a little green goes a long way' in making a city cosmopolitan. Do people of various origins use these places or are they appropriated by specific groups? What kind of coexistence develops amongst the users of these spaces?

Research conducted on modes of coexistence in the public spaces of Montréal's multiethnic neighbourhoods provided some answers to these questions through observing the users of each space, the interaction occurring between them and practices of territorial appropriation or exclusion occurring in them (Germain *et al.* 1995; Germain 1999). Most of these spaces were frequented by people from a wide range of ethno-cultural backgrounds and few of them were appropriated exclusively by any one ethnic group. The dominant mode of public sociability in these spaces can be characterized as an essentially peaceful but distant cohabitation. Little inter-group mixing occurred between different people using these spaces, as people tended to interact mainly with others of their own ethnic origin, age or generation and gender. Ethno-cultural differences seem therefore to be accepted as part of the urban landscape and of daily life in these neighbourhoods.

For theorists like Richard Sennett, spaces need to foster 'meaningful' interaction between members of different cultures to be truly cosmopolitan – the risk being that if contact in public spaces is limited, and ethnic difference is managed 'by principles of non-interaction', then difference will paradoxically breed indifference (Sennett 2002: 47). But we would question, along with Leonie Sandercock, '[w]hy . . . sharing a common destiny in the city necessitates more than a willingness to live with difference in the manner of respectful difference' (2003: 6). Georg Simmel (1979 [1903]) showed long ago that social distance is a defining characteristic of big cities. There is no reason why residents of multiethnic neighbourhoods should be any more demonstrably friendly or involved with each other than those in ethnically homogeneous neighbourhoods, where sociability can be equally superficial and delicately balanced at a 'friendly distance', as in the small and very white English town studied by Crow *et al.* (2002). Moreover, these authors point out that closer, stronger or deeper

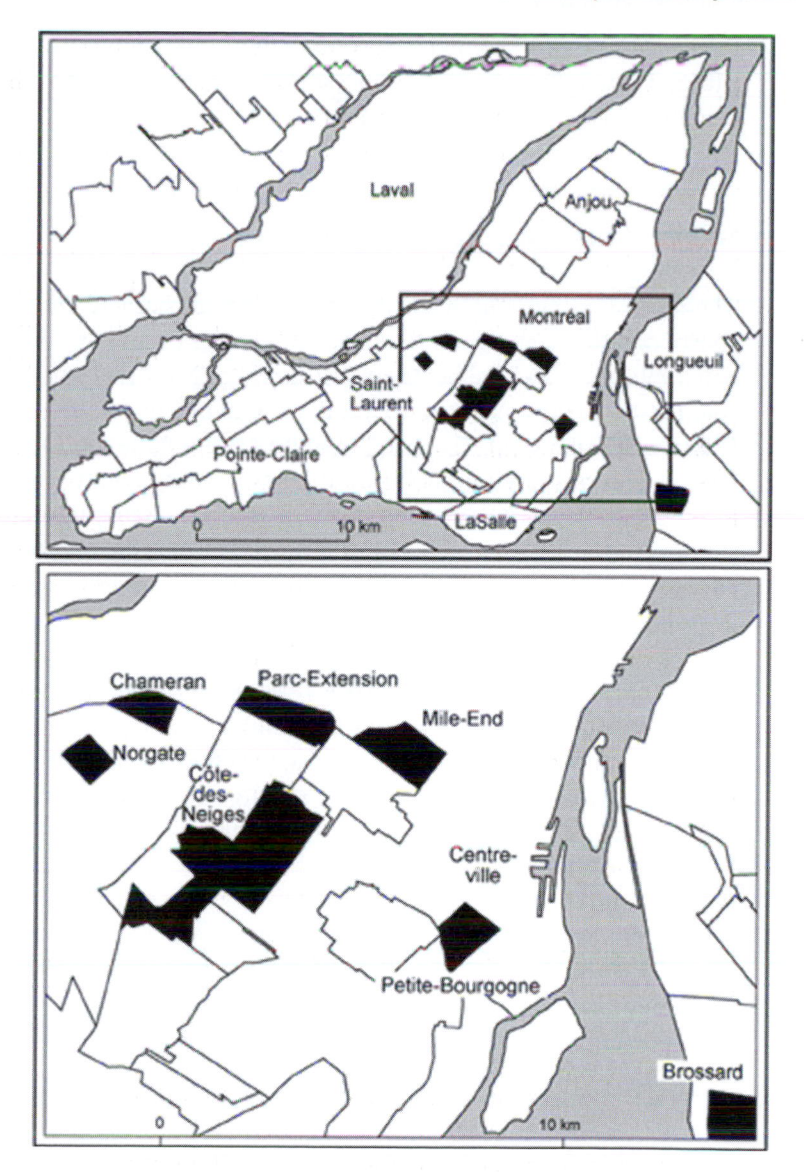

Figure 6.2 ***The seven multiethnic neighbourhoods selected for our study.***
Source: INRS – Urbanisation, Culture et Société.

relationships in the immediate neighbourhood are not necessarily desirable, especially if they are felt to be imposed.

Local civil society in each of the neighbourhoods, and in particular participation by members of ethnic minorities in community organizations, were also investigated. While the peaceful but distant modus vivendi in public spaces reigned in all neighbourhoods, one of them – Mile End – seemed to have a dynamic in community activities with a distinctly cosmopolitan flavour (Rose 1995), and this neighbourhood is considered in more depth below.

Cosmopolitanism as a shared representation of the city

The idea of cosmopolitanism seems not only to characterize but also to consolidate inter-group relations in Mile End. For a long time, this neighbourhood was the northern extremity of Montréal's main corridor of immigrant reception and settlement running along Boulevard St-Laurent. Even today, it boasts a population that is highly diverse in terms of ethno-cultural origins and what might loosely be termed 'lifestyles'. After losing much of its original population to the suburbs in the 1960s, it underwent a renaissance at the end of the 1970s with the arrival of young non-immigrant francophones and anglophones attracted by low rents. Unlike the traditional population, these new arrivals were highly educated and sometimes affluent, and their arrival led to a partial gentrification of the neighbourhood. They settled down beside the original inhabitants, mostly homeowners and tenants from various European and Central and South American countries (as well as a small Chinese community). These include Portuguese families, for example, whose older members have little formal education and who may have been renting the same dwelling for years, not having amassed the means to move out to the suburbs. Then there are the Hassidic Jewish residents, members of an ultra-orthodox sect who arrived in the 1950s, just as an earlier wave of Jewish immigrants was flowing out of the city to the suburbs. The ultra-orthodox community has also lost some members to suburbia, but it is still growing due to natural increase and immigration from New York. What could all these groups possibly have in common?

Cosmopolitanism has, over time, become a trademark of Mile End. Local residents value the everyday sociability of heavily frequented commercial streets and small 'ethnic' shops, where signs proudly declare 'Eight languages spoken here'. On St-Jean-Baptiste Day (24 June), Québec's public holiday and traditionally a very French-Canadian event, the Mile End Citizens' Committee coordinates a convivial multiethnic festival that attracts non-residents as well as residents of almost all backgrounds (though the Hassidim tend to pass through rather than participate). (The festival recently fell victim to its own success, attracting so many non-residents that it became unmanageable, so the Citizens' Committee did not organize it in 2004.) Again, this cosmopolitanism can be criticized as superficial: because it does not involve 'real' intercultural exchange and interaction, it does not go beyond what Latouche (1990) calls 'the cosmopolitanism of the bazaar'. He argues that 'A cosmopolitanism that is closed in on its own discourse . . . leads inevitably to the mere over-consumption of the apparent signs of coexistence and ethnicity' (1997: 13, our translation). But we think that the local urban bazaar has value beyond its commodities. Greek grocers say their stores are still used by ex-residents who have moved to the suburbs; a community worker reports that Latin American immigrants move to the area because they can find the food they like there. In Mile End, 'one can see Montréalers of Hassidic Jewish, Italian, Latin American, Jamaican and old stock Québécois origins, low-income families and members of the new middle classes bitten by the "ethnic" cuisine bug all shopping in the same places' (Rose 1995: 95, our translation). Speciality grocery stores are places where cultures do intersect. Even if the interactions are superficial, difference is rendered more familiar and acceptance of the Other is fostered.

Mile End's cosmopolitanism also shapes how residents deal with local problems. In 1982 the Comité des Citoyennes et des Citoyens du Mile End (Mile End Citizens' Committee) was founded, mostly from among the new, gentrifying residents. Initially preoccupied with aesthetics and traffic calming, it soon began to mobilize on social issues such as getting a new library and reclaiming patches of spare land for much-needed parks. The committee tries to campaign in such a way as to include (and preferably unite) all groups in the neighbourhood. For instance, a meeting we observed included Greek, Portuguese and Hassidic Jewish men and women, amongst others, and multilingual participants provided informal simultaneous translation, which may seem banal but is quite rare in Montréal community politics. Some campaigns are particularly bound up with the neighbourhood's multiethnicity. When the city of Montréal announced in 1986 its intention to rezone certain streets as purely residential, which would have forced several small 'ethnic' shops to close, the Citizens' Committee mobilized a wide range of local residents to oppose the plans (successfully, as it turned out). Residents later fought (but lost) a similar battle in 1992 against the bus lanes that they feared would turn a local high street into a high-speed thoroughfare. Both campaigns testify to the importance of the 'bazaar' to everyday sociability and community-building capacity.

Cosmopolitanism has thus become a shared representation in the neighbourhood of Mile End, a vision that almost everyone agrees with, because the cosmopolitan label is a 'fuzzy' one that can be tailored to suit every resident – from Orthodox Jews to Portuguese elders to young gentrifiers – in one way or another. Some residents feel that the multiethnicity of the neighbourhood is romanticized, masking each group's indifference to the others. And interethnic tensions do arise on occasion. Different norms of neighbourliness can mean that southern European immigrant residents feel that their Franco-Québécois neighbours are uncivil towards them, while the latter in turn can feel the former are oversolicitous. More concrete conflict was provoked by proposals to extend a Hassidic synagogue, but residents reached a 'liveable' resolution after representatives of the various interested parties (including the synagogue's immediate neighbours) engaged in pragmatic negotiation with each other over traffic, parking, noise and suchlike. This outcome contrasted sharply with a similar controversy over a synagogue extension in an adjacent neighbourhood, where the conflict was taken immediately to a tribunal and escalated from thereon in (Germain and Gagnon 2003). In Mile End, conflicts seem to be contained by the commonly held acknowledgement that the Other has a right to be there.

The cosmopolitan project

Might a similar dynamic be understood to characterize all of Montréal? Sometimes a part of a city can stand for the whole. Montréalers can be attached to ethno-culturally diverse neighbourhoods even if they do not live in them, and might cite the visible diversity of, say, Mile End's St-Viateur Street as typical of the entire city (Radice 2000: 96ff.). Bonnie Menes Kahn insists that:

> the cosmopolitan city . . . is one where diversity has created a temporary tolerance, a thriving exchange among strangers. And the project of the place, by force, by design, by chance or coercion, the project is an attempt to benefit from the presence of newcomers and outsiders.
>
> (Kahn 1987: 3)

The challenge is to locate a city's 'project' at all. What are its goals, who might benefit from it and in which sphere does it operate? Montréal's vocation was once to be the metropolis of Canada, but it now vacillates between being a mere metropolis of (French) Québec and a (bi- or multilingual) node in the network of global cities (Germain and Rose 2000: 254ff.). However, could it also/instead be cosmopolitan? In order to think about this, we need to explain briefly how Montréal is situated with respect to Canada's well-known policy of 'official multiculturalism'. The tension between cosmopolitanism and nationalism also features here.

Montréal is the metropolis of the province of Québec, whose electorate twice rejected moves towards independence from Canada in the referenda of 1981 and 1995, but only by an extremely narrow margin in 1995. The drive towards independent nationhood is fostered most intensely (although not only) amongst French-speaking Québécois who trace their roots back to Nouvelle-France. In the foundation mythology of Canada, this group was regarded as one of the country's two founding nations (the other being the British). However, from the Franco-Québécois point of view their claim to sovereignty is based specifically on a constructed history of being a colonized and marginalized people rather than an equal partner in the Canadian project. Although Québec is still part of Canada, many Québécois continue to aspire to independence and Québec is often referred to from within as a nation. For instance, St-Jean-Baptiste Day is known as 'la fête nationale' and the city of Québec as 'la capitale nationale'.

Immigration to Canada from other countries – widely believed to be necessary for keeping the economy and the birthrate afloat – has, of course, complicated the story. Many French Québécois felt that their claims were deliberately undermined when Prime Minister Pierre Elliot Trudeau introduced the 'mosaic' ideology of official multiculturalism in 1971. Suddenly, they were no longer one of two founding nations but simply one minority among many others. This reaction notwithstanding, Québec too has had to deal with, and moreover encourage, immigration. Québec (like most Canadian provinces) has the authority to select its own immigrants, and the ability to speak French – an essential aspect of Québécois identity – gains extra points for applicants in this process. The government of Québec calls its strategy for the management of diversity 'interculturalism' instead of multiculturalism, which carries quite negative connotations in the province for the reasons outlined above (although Danielle Juteau *et al.* (1998) have demonstrated that the actual policies resulting from interculturalism are quite similar to those driven by multiculturalism).

Québec's ethno-cultural diversity and nine-tenths of its immigrants are concentrated in the Montréal region, mostly on the island. This area thus stands in clear contrast to the rest of Québec which is broadly ethnically homogeneous. Although the nationalist project first emerged from francophone Montréal, the size and significance of the anglophone and immigrant populations limit support for Québec's independence on the island (Gagné and Langlois 2002: 111). Québec nationalism used to be seen as based on ethnic group membership, and many immigrants still feel that they would be less welcome in an independent Québec than in the Canada that accepted them (and that they do not therefore wish to reject). However, activists campaigning for the independence of Québec have made their visions more inclusive of ethno-cultural difference (e.g. Bariteau 1998) as long as immigrants adopt the French language.

Interestingly, it could be argued that the instrument used to ensure the survival of French, although assimilationist in its intentions, has turned out to be cosmopolitan in effect. Bill 101 became law in Québec in 1977, consolidating the position of French as the only official language in the province (since 1974). Its greatest long-term impact was to oblige all schoolchildren to attend French-language schools, unless one of their parents had been educated in English, in Canada. The children of immigrants must therefore be taught in French. Immigrants had previously favoured English-language schools, firstly because English was considered to be more conducive to socio-economic mobility, and secondly because the school system used to be based on religious affiliation. French-speaking Catholic schools generally closed their doors to children who were not Roman Catholics (e.g. Jews, Orthodox Greeks), who went instead to Protestant (English-speaking) schools where they did not mix with the majority ethnic group (McAndrew 2003). Bill 101 has therefore forced the majority to share with minorities a crucial (and obligatory) institution of socialization, the (francophone) state school system. Amin (2002) has argued that the greatest potential for negotiating difference in a multiethnic society does not lie in the 'optional' public spaces that people frequent by choice, from which 'the marginalised and the prejudiced [still] stay away', but rather in the 'micropublics' such as schools, workplaces and sports clubs where '"prosaic negotiations" are compulsory' (Amin 2002: 968–9). One would expect that this local 'accommodation' is already taking place in the extremely multiethnic classrooms created in many parts of Montréal by Bill 101 (we would say that it is also happening in Montréal's parks, however optional these spaces are).

Still, the classrooms' cosmopolitanism is an unintentional side-effect of a Bill whose main concern was to reverse the assimilation of immigrants into the English-speaking world. We doubt whether it is appropriate to try to legislate for cosmopolitanism at all, on any political agenda – language policy, nationalism or interculturalism. For instance, cosmopolitanism directly contradicts any principle of *integration* of immigrants, since 'in a stricter sense [it] would entail a greater involvement with a plurality of contrasting cultures to some degree on their own terms' (Hannerz 1996: 103), whereas integration usually implies terms set by the 'host' society. We suggest that the very reason that cosmopolitanism is valuable in Montréal is that it is involuntary, unlegislated: cosmopolitanism by default.

We argue that the debate over Québec's independence to some extent protects Montréal's cosmopolitanism, however this is conceived. Ordinary city-dwellers assume their shared Montréaler identity with relief, even – perhaps especially – when they are politically divided. The political scandals and social tensions that erupt in the city still tend to be provoked by the nationalist question rather than interethnic conflict (although the two are sometimes linked, since nationalists are usually Franco-Québécois and members of ethnic minorities are not often nationalists). It would be naïve to say that police racism or minority underemployment, for instance, do not exist, but rightly or wrongly they are not causes that catch on in the media or are taken up by politicians: *la question nationale* leaves room for little else on the political agenda. Québec, unlike Britain or France, has not seen a publicly fascist municipal councillor or a 'race riot' for decades. Since interethnic co-presence is largely unremarkable, a kind of cosmopolitan consensus can be consolidated at the everyday neighbourhood level.

Montréal is an island city, convinced of its own uniqueness, many of whose inhabitants define their allegiance to Montréal first and foremost, and then in relation to other cities (Helly and van Schendel 2001: 201; Radice 2000: 88ff.). As in Caglar's analysis of the media targeting of young German Turks:

> the sense of belonging is not envisaged as being to nations but to urban spaces. The relationship between the nation and the city is uncoupled. The urban spaces become the strategic arena for the development and reformulation of emergent forms of belonging.
>
> (Caglar 2002: 186)

Many Montréalers' allegiances stretch far beyond the boundaries of Québec and/or Canada, and it is easy for them to symbolically assume their literal insularity vis-à-vis the nation. In this sense, it seems likely that there is indeed a shared representation of Montréal as a cosmopolitan space where one engages, like it or not, with others.

Conclusion

Is Montréal a cosmopolitan city? The answer depends, of course, on one's definition of cosmopolitanism. Montréal is cosmopolitan in terms of its ethno-cultural diversity, its relatively easy economic integration of immigrants, and the openness of its public spaces – parks, squares, commercial streets – to all. However, in terms of its 'project' it is cosmopolitan by default, having become so largely without the aid of municipal policies and at a certain distance from provincial and federal ones. Ad hoc management of diversity such as observed in Montréal (Germain *et al.* 2003) is not necessarily a bad thing. As Amin puts it, 'Mixed neighbourhoods need to be accepted as the spatially open, culturally heterogeneous, and socially variegated spaces that they are, not imagined as future cohesive or integrated communities' (2002: 972).

Cosmopolitanism – understood idealistically rather than cynically – might therefore be ill served by official policies, and might best be understood as a kind of anti-model. Rather than enforcing the city's residents into a project of citizenship, cosmopolitanism in Montréal reflects the organic reality of the city and lay uses of the term. It is a shared representation, a metonymic projection of the valued diversity of some neighbourhoods onto the city as a whole, which in turn influences how people treat each other. Montréal's cosmopolitanism might, after all, simply consist of the right not to be defined by others as belonging to a single or particular community. This 'fuzzy' cosmopolitanism by default is invariably composed, to some degree, of consumer cosmopolitanism. Enjoying superficial relations with one's diverse neighbours or exploring novel purchases from different immigrant groups' grocery stores *is* about knowing and consuming Others, without always making the effort to understand them first. However, perhaps it is a necessary basis for the intercultural dialogue that constitutes harmonious coexistence in the city.

References

Amin, A. (2002) 'Ethnicity and the multicultural city: living with diversity', *Environment and Planning A*, 34: 959–80.

Bariteau, C. (1998) *Québec, 18 Septembre 2001*, Montréal: Éditions Québec Amérique.

Binnie, J. and Skeggs, B. (2004) 'Cosmopolitan knowledge and the production and consumption of sexualized space: Manchester's gay village', *Sociological Review*, 52: 39–61.

Caglar, A. (2002) 'Media corporatism and cosmopolitanism', in S. Vertovec and R. Cohen (eds), *Conceiving Cosmopolitanism: Theory, Context, and Practice*, Oxford: Oxford University Press.

Cheah, P. (1998) 'Introduction Part II: The cosmopolitical – today', in P. Cheah and B. Robbins (eds), *Cosmopolitics: Thinking and Feeling Beyond the Nation*, Minneapolis, MN: University of Minnesota Press.

Citizenship and Immigration Canada (2002) *Facts and Figures 2001: Immigration Overview*, Ottawa: Citizenship and Immigration Canada, Strategic Policy, Planning and Research. Available at www.cic.gc.ca/english/pdf/pub/facts2001.pdf (accessed 10 September 2004).

Crow, G., Allan, G. and Summers, M. (2002) 'Neither busybodies nor nobodies: managing proximity and distance in neighbourly relations', *Sociology*, 36: 127–45.

Debarbieux, B. and Perraton, C. (1998) 'Le parc, la norme et l'usage: Le parc du Mont Royal et l'expression de la pluralité des cultures à Montréal', *Géographie et Cultures*, 26: 109–27.

Department of Justice Canada (2004) *Employment Equity Act 1995, c.44.* Available at http://laws.justice.gc.ca/en/E-5.401/text.html (accessed 10 September 2004).

Gagné, G. and Langlois, S. (2002) *Les raisons fortes: nature et signification de l'appui à la souveraineté du Québec*, Montréal: Les Presses de l'Université de Montréal.

Germain, A. (1999) 'Les quartiers multiethniques montréalais: une lecture urbaine', *Recherches Sociographiques*, 40: 9–32.

Germain, A. and Gagnon, J. E. (2003) 'Minority places of worship and zoning dilemmas in Montréal', *Planning Theory and Practice*, 4: 295–318.

Germain, A. and Rose, D. (2000) *Montréal: The Quest for a Metropolis*, Chichester: Wiley.

Germain, A., Archambault, J., Blanc, B., Charbonneau, J., Dansereau, F. and Rose, D. (1995) *Cohabitation interethnique et vie de quartier*, Québec: Gouvernement du Québec, Ministère des Affaires Internationales, de l'Immigration et des Communautés Culturelles, Collection Études et Recherches No. 12.

Germain, A., Dansereau, F., Bernèche, F., Poirier, C., Alain, M. and Gagnon, J. E., with the collaboration of Polo, A.-L., Legrand, C., Vidal, L., Ainouche, L. and Daher, A. (2003) *Les pratiques municipales de la gestion de la diversité à Montréal*, Montréal: INRS – Urbanisation, Culture et Société.

Gouvernement du Québec (2004) *Tableaux sur l'immigration au Québec 1999–2003*, Québec: Ministère des Relations avec les Citoyens et de l'Immigration, Direction de la Population et de la Recherche. Available at www.mrci.gouv.qc.ca/publications/pdf/Immigration_Quebec_1999-2003.pdf (accessed 10 September 2004).

Hannerz, U. (1996) *Transnational Connections: Culture, People, Places*, London: Routledge.
Helly, D. and van Schendel, N. (2001) *Appartenir au Québec: citoyenneté, nation et société civile*, Sainte-Foy: Les Presses de l'Université Laval/Les Éditions de l'IQRC.
Hiebert, D. (2002) 'Cosmopolitanism at the local level: the development of transnational neighbourhoods', in S. Vertovec and R. Cohen (eds), *Conceiving Cosmopolitanism: Theory, Context, and Practice*, Oxford: Oxford University Press.
Jedwab, J. (2003) *Want the Good News or the Bad News First? Unemployment and Visible Minorities in Canada's Cities in 2001*, Montréal: Association for Canadian Studies.
Juteau, D., McAndrew, M. and Pietrantonio, L. (1998) 'Multiculturalism à la Canadian and integration à la Québécoise: transcending their limits', in R. Baubock and J. Rundell (eds), *Blurred Boundaries: Migration, Ethnicity, Citizenship*, Aldershot: Ashgate.
Kahn, B. M. (1987) *Cosmopolitan Culture: The Gilt-Edged Dream of a Tolerant City*, New York: Atheneum.
Latouche, D. (1990) *Le Bazaar: des anciens Canadiens aux nouveaux Québécois*, Montréal: Éditions Boréal.
Latouche, D. (1997) *Mondialisation et cosmopolitisme à Montréal*, Montréal: INRS – Urbanisation, Culture et Société, Cahiers Culture et Ville No. 97-10.
Law, L. (2002) 'Defying disappearance: cosmopolitan public spaces in Hong Kong', *Urban Studies*, 39: 1625–45.
Linteau, P.-A. (2000) *Histoire de Montréal depuis la confédération*, second edn, Montréal: Boréal.
McAndrew, M. (2003) 'Immigration, pluralisme et éducation', in A.-G. Gagnon (ed.), *Québec: état et société*, tome 2, Montréal: Éditions Québec Amérique.
McNicoll, C. (1993) *Montréal: une société multiculturelle*, Paris: Belin.
Radice, M. (2000) *Feeling Comfortable? The Urban Experience of Anglo-Montrealers*, Sainte-Foy: Les Presses de l'Université Laval; trans. M. Radice (2000) as *'Feeling comfortable?' Les Anglo-Montréalais et leur ville*, Sainte-Foy: Les Presses de l'Université Laval.
Remy, J. (1990) 'La ville cosmopolite et la coexistence inter-ethnique', in A. Basteneier and F. Dassetto (eds), *Immigration et nouveaux pluralismes*, Brussels: De Boeck; reprinted in J. Remy (1998) *Sociologie urbaine et rurale: l'espace et l'agir*, Paris: L'Harmattan.
Rose, D. (1995) 'Le Mile-End, un modèle cosmopolite?', in A. Germain *et al.*, *Cohabitation interethnique et vie de quartier*, Québec: Gouvernement du Québec, Ministère des Affaires Internationales, de l'Immigration et des Communautés Culturelles, Collection Études et Recherches No. 12.
Sandercock, L. (2003) 'Rethinking multiculturalism for the 21st Century', Research on Immigration and Integration in the Metropolis Working Paper Series No. 03-14, Vancouver: Vancouver Centre of Excellence.
Sennett, R. (2002) 'Cosmopolitanism and the social experience of cities', in S. Vertovec and R. Cohen (eds), *Conceiving Cosmopolitanism: Theory, Context, and Practice*, Oxford: Oxford University Press.
Simmel, G. (1979 [1903]) 'Métropoles et mentalité', in Y. Grafmeyer and I. Joseph (eds), *L'école de Chicago*, Paris: Champ Urbain.
Statistics Canada (2004) *2001 Community Profiles*. Available at www12.statcan.ca/english/profil01/PlaceSearchForm1.cfm (accessed 10 September 2004).

Vertovec, S. and Cohen, R. (2002) 'Introduction: conceiving cosmopolitanism', in S. Vertovec and R. Cohen (eds), *Conceiving Cosmopolitanism: Theory, Context, and Practice*, Oxford: Oxford University Press.

Ville de Montréal (2000) *Construire ensemble: Orientations 2000–2001–2002 relations interculturelles*, Montréal: Ville de Montréal.

Winock, M. (1997) *Le siècle des intellectuels*, Paris: Éditions du Seuil.

Cosmopolitan camouflage: (post-)gay space in Spitalfields, East London

Gavin Brown

For centuries Spitalfields, in East London, has accommodated successive waves of new immigrants from across the world – Huguenots, East European Jews and, in the last 30 years, Bangladeshis. Today traces of these various immigrant communities live on in the architecture and everyday life of the area. Over the years, Spitalfields has featured heavily in the art of Gilbert and George, and appeared in many works of fiction (Lichtenstein and Sinclair 1999; Ali 2003). It has also been the focus of several academic studies (Jacobs 1996; Pinder 2001). In these ways, Spitalfields has come to be consumed (inter)nationally.

A decade ago, Spitalfields was an area in physical and economic decline, with high rates of unemployment and poor-quality, overcrowded housing stock. The wholesale fruit and vegetable market had closed and the continuing viability of the area's street markets was in question. The only major sources of employment in the area were in the local 'Indian' restaurants on Brick Lane and the sweated garment trade.

Since then, the neighbourhood has been subjected to a series of regeneration programmes (Cityside Regeneration 2002), which have built upon elements of the area's history, to increase visitor numbers to the area, foster it as a hub for the cultural industries and boost the local night-time economy. Although cheap property and the rich history of the area have attracted gentrifiers to Spitalfields since the 1970s, the recent phase of regeneration has significantly accelerated this trend.

This chapter will consider how these different 'Spitalfields' combine to construct the area as a symbolic cosmopolitan zone and to explore who gains and who loses from this process. I will pay particular attention to the ways in which Spitalfields provides contradictory opportunities for non-heterosexual men living and working in the area. I shall consider how non-heterosexual men from various class and ethnic backgrounds engage with cosmopolitan difference and attempt to utilise these spaces of cosmopolitan cultural consumption to mitigate other social exclusions. In doing so, I draw on existing work on the role of gay spaces in the development of cosmopolitan zones in the contemporary metropolis. I also utilise Lefebvre's (1991) thoughts on the production of social space to consider the extent to which social mixing is facilitated by the spatial practices that contribute to the (re)presentation of Spitalfields as a cosmopolitan space.

Sexual difference and cosmopolitanism

Cosmopolitanism has become one of the most desirable forms of cultural capital within late capitalism (Rushbrook 2002). As a result, some cities have incorporated the presence of visible ethnic diversity into their place marketing strategies; and, in the UK, a number of deprived inner city districts have sought to capitalise on local minority ethnic populations as a driver for social and economic regeneration (albeit usually just by highlighting 'ethnic' restaurants and carnivals (Jacobs 1996)). Spitalfields has been subject to both processes, with Brick Lane's Bengali restaurants serving as a focus for the regeneration of the area and the newly regenerated area being promoted internationally as an example of 'multicultural' London.

I should make clear that I see multiculturalism and cosmopolitanism as two quite distinct concepts. Cosmopolitanism is about more than just an openness to ethnic difference. It can encompass a willingness to sample and fuse a broad set of (sub)cultural forms, including sexual difference and bohemian lifestyles. Mitchell (2003) has suggested that the liberal, inclusive multiculturalism developed since the 1970s is giving way to a more individualistic and competitive adoption of 'strategic cosmopolitanism' in the era of deterritorialised neo-liberal capitalism with its reliance on innovation and creativity in the 'knowledge economy'. While I accept Mitchell's argument that the growth of cosmopolitan lifestyles is linked to wide-ranging political economic changes at a global scale, there is a danger that cosmopolitan tastes become over-associated with relatively privileged middle-class fractions. Today working-class people (especially in large metropolitan centres) lead more ‘cosmopolitan’ quotidian lives than ever before – it should be self-evident that it is not just the middle classes who enjoy eating curries and Thai food or drink espressos. Nevertheless, cosmopolitanism remains a distinctly middle-class habitus through which certain middle-class fractions distinguish themselves both from the working classes and from more ‘conservative’ sections of the middle class.

In recent years, an increasing number of cities have promoted the presence of ‘gay villages’ as an additional example of their cosmopolitan chic. As Florida (2002: 256) has suggested, in the context of the United States, ‘to some extent, homosexuality represents the last frontier of diversity in our society, and thus a place that welcomes the gay community welcomes all kinds of people’. Working from similar assumptions, Bishop (2000) has argued that the presence of a visible gay population in a city is more important than simply an additional marker of diversity. With the proclamation ‘where gays go, geeks follow’, he suggests that toleration of gay people becomes a marker of social liberalism in a city that in turn attracts cultural and technological entrepreneurs who can act as a driver for the growth and sustainability of the regional economy.

The dynamics of Spitalfields’ development as a cosmopolitan zone have been slightly different. The concentration of Bengali businesses locally, and the area’s history of receiving successive waves of immigrants, have acted as the main focus of its promotion as an exemplar of cosmopolitan London. Simultaneously, the area has a significant gay presence, supported by a small and relatively diffuse infrastructure of gay venues. Interestingly, though, few of these venues are ever included in place marketing materials or tourist guides about the area. Part of my motivation in writing this chapter has been to explore the place of sexual difference in a city neighbourhood that is promoted as a cosmopolitan space without any explicit acknowledgement of the role of queer bodies in contributing to the area’s cultural mix.

Before going on, I should clarify my use of various terms relating to markers of sexual difference. In this chapter, I use ‘gay’ to refer to those people who identify with mainstream homosexual identities. It also refers to those spaces where these identities dominate and are (re)produced and reified. The term ‘non-heterosexual’ is used as an umbrella term for all men who have sex with men, whether they categorise their identity on this basis or not. 'Post-gay' spaces are those where sexual difference is visible and acknowledged without being the central marker of the space. Finally, 'queer' spaces and identities are those that consciously disrupt normative sexual and gender binaries. Despite this attempt to distinguish and define these categories, I should stress that I do not view them as discretely bounded; rather, they are overlapping, relational and place-specific. What is (homo)normatively 'gay' in the context of cosmopolitan inner London might be deeply 'queer' ten miles away in the suburbs.

Binnie and Skeggs (2004: 39) have suggested that studies of cosmopolitanism have tended to avoid considerations of the mutually constitutive politics of sexuality and class in favour of discussions of the intersections of ethnicity with other social categories. Like them, I want to put class back on the agenda, alongside sexuality and ethnicity. However, where they focused their discussion on the interplay of class and sexuality as they are expressed in the gay spaces of Manchester's Gay Village, I want to explore how sexuality fits into the social dynamics of an area whose cosmopolitan credentials are defined primarily by differently classed appropriations of minority ethnic culture.

In the context of Manchester's Canal Street district, Binnie and Skeggs (2004) have argued that the promotion of the area as a cosmopolitan space allows heterosexual people (and women in particular) to overcome their discomfort with being 'out of place' in gay space. As they suggest,

> the term cosmopolitan is useful in helping to understand the unease and discomfort with being an appropriate or 'proper' user of space which requires a fixity of identity, a possession of the right personae to pass through and occupy the space . . . We argue that behind and within the articulation and desire for the fluidity of identity associated with the use of the term cosmopolitan, the rigidities of class and lesbian and gay identities are reproduced. In particular, class entitlement plays a major role in articulating *and* enabling who can be included and excluded from this space.
>
> (Binnie and Skeggs 2004: 40, emphasis in original)

Similar processes and tensions are at work in Spitalfields. However, because the area's main claim to being 'cosmopolitan space' rests on the presence of a large local Bengali population, alongside artists and other cultural producers, the dynamics are quite different. In Spitalfields, the boundaries of class and ethnic identities are reinforced and reproduced, but sexual identities can, at times, become more fluid and less easily defined.

It has frequently been argued that the performance of cosmopolitanism relies on the deployment of a certain kind of cultural capital. To be cosmopolitan one must be open to encounters with the 'Other' and be willing and able to appropriate something of that 'otherness' (Hannerz 1992). It is also about generating a certain kind of authority from these appropriations (Binnie and Skeggs 2004: 42). All of which raises questions about how one accesses the skills, knowledge and cultural capital in order to be in a position to develop a cosmopolitan disposition out of intercultural encounters. These are important questions, but there is a danger that in answering them we (re)produce an exclusionary definition of 'cosmopolitanism' as the preserve of the educated, (mostly) white, urban middle classes. To do so ignores and invalidates the multiple hybrid identities constructed and performed by many working-class youths, particularly those from minority ethnic communities, as they attempt to make sense of the numerous conflicting pressures on their lives.

Such considerations are particularly important when thinking about the multitude of overlapping spatial practices that produce Spitalfields as a cosmopolitan space. If cosmopolitanism is about a certain kind of openness and curiosity about different cultures, often constituted through consumption, in which multiple identities and lifestyles are tried on for size and then discarded (Rushbrook 2002), then in Spitalfields the social and spatial practices of young Bengalis must be factored into the equation as much as those of the 'fashionistas' browsing the designer boutiques or the gaggles of City workers out for a curry.

Whilst discussions of cosmopolitanism tend to focus on the appropriation of ethnic difference, Rushbrook (2002: 188–9) has suggested that 'queer space is one more place in which cultural capital can be displayed by the ability to negotiate different identities, to be at ease in multiple milieus, to manoeuvre in exoticized surroundings'. She is primarily talking about the consumption of gay space by white heterosexual tourists. However, in Spitalfields, the use of (post-)gay and queer space can be a means by which young, non-heterosexual Bengali men accumulate cosmopolitan cultural capital through open encounters with men of differing social backgrounds. After all, on streets around Brick Lane, where it is their 'own' culture that is reified through the consumption practices of other Londoners and tourists, the pressures to perform a certain set of 'acceptable' identities are far greater and the opportunities to actively engage with difference are reduced.

Whenever urban space is consciously promoted as cosmopolitan, the space becomes sanitised in order to maximise its appeal as a desirable site of consumption for the broadest possible audience. This is particularly the case when gay space is branded as 'cosmopolitan', as Binnie and Skeggs (2004: 47) have highlighted: 'the more threatening, less easily assimilated aspects of urban sexual dissidence are rendered invisible – and most specifically, the sexual side of gay men's urban cultures are downplayed, with only certain aspects of gay male culture promoted'. This can be seen, to an exaggerated degree, in the promotion of Spitalfields as a cosmopolitan leisure zone, where the presence of a number of gay venues barely registers in the area's place marketing strategies and the location of Britain's largest gay sauna is kept studiously under wraps (Shoreditch Map Company 2004).

Despite this rendering of gay space as invisible, I do not want to suggest that the promotion of Spitalfields as a space for middle-class cosmopolitan consumption has necessarily been a uniformly bad thing for (some) non-heterosexual people who live, work and play in the area. Like Binnie and Skeggs (2004: 44), I acknowledge the importance of examining the sexual, class and ethnic politics of cosmopolitanism together, to avoid the common practice of subsuming one set of politics within the others. Looking at the everyday spatial practices of Spitalfields' cosmopolitan venues and street life through a queer lens prompts me to question whether the invisibility of gay venues and non-commercial queer space in the area might actually be of benefit to some sexual dissidents, just as others find new opportunities in the more mixed leisure spaces that the area has on offer.

I want to suggest that, to some extent, the dynamics of everyday life in Spitalfields contributes to the suspicion that, for certain sections of the population at least, 'gay' identities as they have been defined and lived over the last 35 years have begun to outlive their usefulness. Sinfield (1998) has put forward a case that 'post-Stonewall' lesbian and gay identities are tied to the metropolitan centres of capital in the postcolonial world; and, as a result, exclude as many people as they include. As he has elaborated:

> People belonging to racial minorities, queers and people who term themselves 'bisexual' are still, for the most part, declaring some kind of relation to current lesbian and gay concepts and institutions . . . [However,

> a] fourth disturbance in metropolitan lesbian and gay images comes from people who scarcely see themselves in relation to prevailing modes of sexual dissidence: men who have sex with men, and women who have sex with women, while living generally as heterosexuals and regarding themselves as basically 'normal'.
>
> (Sinfield 1998: 11)

A related shift in contemporary conceptualisations of masculinity has been observed by Simpson (2002), who coined the phrase 'metrosexual' to satirise a particular layer of young urban men (of all sexualities) who are turning their backs on more traditional expressions of masculinity. Compared to the 'old-fashioned' heterosexual model of masculinity, the metrosexual is 'less certain of his identity, less altruistic, more interested in his image – programmed to consume' (Simpson 2002: 143–4). The streets of Spitalfields, Hoxton and Shoreditch are awash with 'metrosexuals'. After all, it was David Beckham, the metrosexual par excellence, who popularised the ubiquitous late 1990s hairstyle previously known as the 'Hoxton fin'. Indeed, it is to the emerging fashion styles worn by the art students, DJs and web designers who hang out in the area that the fashion industry looks for inspiration that can be sold on to those men who aspire to embody these new expressions of masculinity.

Perhaps, if the presence of gay space is downplayed in Spitalfields, it is not simply an attempt to render unruly, hyper-sexualised spaces invisible, but an indication that, for certain social layers who enjoy this cosmopolitan playground, mainstream 'gay' identities have outlived their usefulness. To this end, this chapter will consider the extent to which the production of Spitalfields as a cosmopolitan zone has also produced a layer of 'post-gay' space in which the need to clearly define and delineate our sexualities is largely deemed unnecessary.

Nevertheless, alongside this (largely) optimistic line of enquiry, I find that it is still important to question whether these emerging consumption practices 'inscribe new or reinforce current exclusionary practices along the lines of race, ethnicity, class and gender' (Rushbrook 2002: 184). In particular, I shall question the extent to which white gay consumers act voyeuristically and contribute to the exoticisation and marginalisation of (sections of) the local Bengali community by participating in the commodified cosmopolitan whirl that Spitalfields has become. With this in mind, in the next section of this chapter I review Lefebvre's work on the city to consider how his ideas can be employed as a framework for the analysis of everyday life in Spitalfields.

The production of (cosmopolitan) space

In *The Production of Space*, Lefebvre (1991) set about tracing the inner dynamics of social space under capitalism. He proposed that there are three 'moments' in the production of social space – representations of space, representational space and spatial practices. First there are 'representations of space', that is the space conceived by urban planners and other bureaucrats. These representations of space are always abstract because they are conceived and constructed through discourse. They are not encountered through daily life, but through the plans, maps and planning codes that shape how space is conceptually ordered.

I start my analysis with an examination of how city planners (primarily employed by the local regeneration agency) have conceived and reshaped the physical environment of Spitalfields and, in the process, imposed a new symbolic meaning on the area. When Cityside Regeneration (2002: 5) began work in 1997 they identified a clear need for regeneration, including 'one of the largest concentrations of buildings of architectural interest at risk in London' and 'limited recognition of the potential of the most multi-cultural community in Europe' (Cityside Regeneration 2002: 5). From the outset of their work, one of Cityside's three main strategies for regenerating the Spitalfields was 'developing visitor attractions, improving the environment and supporting events, all designed to increase the number of visitors to the area who will spend money within it' (Cityside Regeneration 2002: 5).

This strategic objective has been implemented in various ways. The strategy has had a significant impact on the physical appearance and uses of the local built environment. New street furniture, in the form of an ornamental arch and 'Bengali-style' street lamps, has been installed on Brick Lane to consolidate its identity as an ethnic enclave. Financial assistance and marketing advice has been provided to businesses around Brick Lane 'to improve their displays and branding' (Cityside Regeneration 2002: 6). In this way, the area has been rebranded as 'Banglatown'. Initially this was part of a tourism strategy designed to boost the local economy by promoting the culinary delights of the many restaurants around Brick Lane. But Banglatown now also exists as a political entity following the renaming of the local electoral ward. In addition to their efforts to raise the profile of Brick Lane as an appealing, tourist-friendly ethnic neighbourhood, Cityside also attempted to develop the area as an important hub of cultural production. To this end, they enabled young designers to set up business in the area. As part of this strategy, the former Trumans' Brewery on Brick Lane has been brought back into use as a complex of bars, boutiques, gallery space and design workshops.

For Cityside, the creation of Banglatown and their support for the creative industries were not seen as separate strands of work, but a conscious effort to foster a multifaceted cosmopolitan zone in the area. This can be seen from Cityside Regeneration's (2002: 7) evaluation of their 'Raising the Profile' programme, which they claim, in self-congratulatory terms, has 'led to the successful creation of annual events such as Alternative Fashion Week, Baishakhi Mela, Brick Lane Festival and many other community events which draw increasing numbers of visitors to the area'.

On the back of these various planning initiatives, the physical shape of the local area has changed – alongside the 'ethnic' street furniture, upgraded restaurant frontages and new retail outlets, the built environment of Spitalfields has also been reshaped by numerous loft conversions and new-build 'luxury' apartment blocks, built to accommodate the increasing numbers of affluent professionals who have been attracted to the area for its increasingly lively night-time economy and the cultural capital that can be accrued by living in such a 'cosmopolitan' area.

These regeneration strategies have been 'successful' in another way too: by building on existing elements of the area, they appear less consciously planned. By boosting existing trends in the area, the regeneration initiative has succeeded in 'naturalising' itself. By playing to the history of the area, the regeneration of Spitalfields has also erased the most contested aspects of local history and rendered the redevelopment ahistorical. In this way, the middle-class professionals who gain most from this new cosmopolitan space can ignore the contradictions around them and feel comfortable with their privilege.

Of course, the extent to which these regeneration initiatives have really made any positive material difference to the lives of local Bengali residents is open to question – not least because the (newly renamed) Banglatown ward remains one of the poorest urban neighbourhoods in Western Europe. Only now the area also has many more affluent middle-class residents and is marked by even greater social polarisation.

The second moment in Lefebvre's triad is 'representational space', which he describes as a vernacular, affective layer in the social landscape that 'overlay[s] physical space, making symbolic use of its objects' (1991: 39). It is space as it is experienced through the symbols and images perceived by its users and inhabitants as they draw on physical objects in the landscape, and representations of that landscape produced by artists and writers, in order to symbolise lived experience and give it meaning. It is space that lacks cohesiveness and does not obey the rules. It can, thus, produce counter-discourses and open up possibilities to think differently about space. There is no space in this chapter to explicitly consider how social space in Spitalfields is perceived through artistic representations, but I draw readers' attention to the works of literature referred to in my introduction (Lichtenstein and Sinclair 1999; Ali 2003) and the experiential artworks discussed by Pinder (2001). Some of the more quotidian perceptions of local social space by its inhabitants are discussed in the section that follows on the everyday spatial practices of cosmopolitan life in the area.

Finally, there are 'spatial practices' – the everyday routines and experiences that 'secrete' their own social space. These spatial practices directly link and mediate between people's experience of authoritarian, bureaucratically conceived space and the sites where they live out the relative freedom of their everyday lives. They provide some kind of societal cohesion whilst also potentially exposing the contradictions inherent to capitalist society. For example, town planners may designate certain uses for a given city street or neighbourhood. But media representations may contest the reality of that plan and lead users of that space to experience it in a way other than was intended.

Spatial practices of cosmopolitan everyday life

The Spitalfields area, or at least parts of it, at certain times of the day and week, is today a relatively safe comfortable space for gay people to use. At the weekend it is a common sight to see (invariably white) young lesbian and gay couples walking hand in hand through the street markets. But there are limits to this safety, and just a few blocks away, off the main commercial streets, visible expressions of affection between same sex lovers brings the risk of violence (*East London Advertiser* 2004).

In this section, I will examine the everyday spatial practices by which non-heterosexual men use different local sites to their own advantage.

Rather than examine how and why these men use the small number of gay venues on the fringes of the Spitalfields area, I shall concentrate my discussion on their use of a number of 'mixed' venues and public spaces. The first cluster of sites, a number of bars and coffee shops on Brick Lane, could be described as 'post-gay' spaces (Sinfield 1998) in that amongst their clientele are many people who identify themselves as 'lesbian', 'gay' or 'bisexual', but the sites themselves do not demand the assertion of one identity or another. Most times they contain a majority of heterosexuals, less frequently gay people may be in the majority, but that is not necessarily important. In these sites there is an openness and acceptance of (sexual) difference. The final site, a public toilet in which men meet each other for sex, is better described as 'queer space' because there actions speak louder than words and identity categories can become blurred.

Brick Lane is home to several independent coffee shops. They fulfil many of the functions one would expect of coffee shops in a mixed-use inner-city neighbourhood – a stop-off point for a quick hit of caffeine on the way to work and a hang-out joint for shifting groups of friends. The staff are mostly artists and students, many of them from countries other than Britain. Some of them are gay men, but many more of the male 'baristas' are of more indeterminate sexuality – the classic 'metrosexuals'. The customers are equally diverse and include a large number of gay folk. None of these venues tend to draw many customers from the local Bengali population.

Mostly, the gay men and women who frequent these coffee shops tend to reflect the broader demographics of the clientele. They may be from many corners of the world, but they are mostly cultural producers or young professionals, and mostly white. They use the coffee shops for the same range of purposes as the other customers – a young new media designer holds his business meetings there, groups of young gay men chill out together on a Sunday afternoon after a night out clubbing. Even the few non-white gay men who frequent the place reek of privilege – I recently overheard two wealthy young men (one from Pakistan, the other from Singapore) discussing how they were deliberately prolonging their university studies so that they could continue to live an openly gay lifestyle in London before returning 'home' to confront their parents' expectations of marriage and grandchildren.

At one point, a couple of years ago, a disproportionate number of the (few) Bengali men visiting the coffee shops were also gay. Unlike the local gay bars, the coffee shops carry no particular stigma for men from Muslim communities. Equally, in a venue whose regular customers are drawn from so many nationalities and every sexuality, few people (other than the beady-eyed queer geographer in the corner) are likely to give a second glance to a group of young men sharing their latest (homo)sexual adventures or a thirty-something Bengali man whispering sweet nothings to his (white) lover. Such episodes simply do not stand out as unusual or worthy of any more comment than the discussions taking place simultaneously at every other table in the place. I mention this here to reinforce my point that, for the most 'cosmopolitan' inhabitants of Spitalfields, cultures of sexual difference are included on the menu of divergent experiences they are open to exploring and accepting; but also that this tolerant atmosphere can provide cover for *men* from less privileged social groups. More recently, however, as I have come to write up this chapter for publication, the regularity with which I have noticed non-heterosexual Bengali men using these sites has diminished. It may be that the 'cosmopolitan' spaces of Brick Lane have finally been consolidated as the preserve of the fashionable 'bohemian' set and are no longer a comfortable retreat for even the most daring local Bengali men.

Located in the former Trumans' brewery on Brick Lane, the Vibe Bar was one of the first of the new breed of bars to open in the area. Time and again, its name gets mentioned to me in conversation as a 'mixed' bar that is favoured by local gay men. It also gets listed with considerable regularity in the 'favourite bar' section of local gay men's personal profiles on various internet 'dating' services. There would appear to be a number of different (if overlapping) reasons for this. For some men living or working in the immediate Spitalfields area, it is simply a convenient and comfortable local bar, albeit one that signifies a certain level of attachment to the cosmopolitan and bohemian image of the area. However, a disproportionate number of the men listing it as one of their favourite bars on gay internet sites appear to define their sexuality as 'bisexual' or 'open-minded' rather than 'gay'. Gary, a support worker in a local hostel for the homeless, describes why he likes the bar:

> I use quite a lot of mixed venues now, like the Vibe Bar, well it's mainly the Vibe Bar . . . It's just a mixed venue and I like the area, it's in Brick Lane and I work in that area. Yeah, it's just nice. It's not so concentratedly gay . . . [It's] a lot less cruisy, well there is cruising that goes on there, but it's a lot more subtle. It's just not so in your face, I suppose.
>
> (personal interview)

Gary's comment that subtle homoerotic cruising takes place in the Vibe Bar reinforces my assertion that it is used by a significant number of non-heterosexual men, who are put off by the more 'in your face' atmosphere on most of the gay scene.

The final site that I want to consider in this section is a public toilet (or 'cottage') near Brick Lane. The men who cruise this site belong to many different class fractions, determined partly by their occupation and the economic sector within which they are employed, but also by the ways in which class is 'over-determined' by other social relations.

Over recent years, I have met a very diverse group of men at the cottage. For each of them, class, ethnicity and sexuality work together in very distinct and, at times, contradictory ways. Mixing with the many (British) nurses, bankers and IT workers that use the cottage have been a former rising star of Latin American banking, now working as a hospital domestic in London and dreaming of 'married bliss' with a British gay professional; a Lithuanian law graduate working as a *sans papier* security guard (who swears blind that he has a girlfriend back home); and a trio of Bengali teenagers who adopt a menacing 'rude boy' swagger on the street that mutates into camp frivolity the moment they step inside the toilet.

For some of these men the insecurities of their position as migrant labourers and the painful separation from their friends, family and 'home' communities may be partly compensated by a greater freedom to exhibit and explore 'taboo' sexual desires and identities. At a more 'local' level, the cottage plays a similar function for several of the Bengali men and those from London's other more established minority ethnic communities. For these marginalised immigrant and minimum wage workers, the adoption of a gay identity can (despite their dreams and desires) be fairly low down on the list of priorities. In the cottage, class, ethnicity and nationality work together in complex ways and, as Hennessy (2000: 109) has stressed recently, although new 'postmodern' sexual identities are beginning to travel beyond the major imperialist economies, wealth and education seem to be a prerequisite for their adoption in less developed countries.

Public sex sites can serve to throw reified identities into confusion – a man does not have to have adopted a gay identity in order to join in the action and these sites throw together a far greater diversity of men than the ever expanding array of more specialised niche markets on the commercial gay scene. They also create opportunities for men to assert and articulate their homoerotic desires in ways that may be impossible to achieve in other spaces of their everyday lives. Delany (1999) has argued that this spontaneous, un-commodified erotic contact between strangers is the basis of (queer) urban life. Of course, as Amin (2002: 969) has highlighted, habitual contact is no guarantor of cultural exchange, in itself, and can often reinforce group identities. However, in the context of the cottage, men find themselves outside their quotidian routines and in 'moments of cultural destabilisation' (Amin 2002: 970) that offer the possibility of engaging in a common activity with strangers (in this case, the pursuit of sexual release) that challenges preconceived notions of 'self' and 'Other'. In this sense, I believe that the sites discussed here help establish a dialogue between different subject positions and also create space where more marginalised, non-heterosexual men (especially men of colour) can articulate their desires and explore new or alternative identities.

Hannerz has identified a position of 'in-betweenness' (1992: 200) as being a central marker of the cosmopolitan, caught between two or more worlds. In the sites explored above, it is also often the place of non-heterosexual men of colour and those who find themselves in the most marginal positions within the British economy. While this can be a precarious position to find oneself in, it does suggest the need for a more nuanced reading of the place of non-heterosexual Bengali men and more recent immigrants in relation to the on-going gentrification and commodification of Spitalfields as a cosmopolitan leisure zone.

Unresolved contradictions

In this chapter, I have considered the extent to which the invisibility of gay venues and non-commercial queer space within the context of cosmopolitan Spitalfields might actually be of benefit to some non-heterosexual men. In Spitalfields, local government regeneration strategies have largely erased the social struggles that have shaped the area over recent centuries, whilst they have also attempted to draw on this history to inspire and justify their plans for the (re)presentation of the neighbourhood as a cosmopolitan leisure zone that can both attract more tourist spending to the area and provide the necessary infrastructure to consolidate the locality as a hub for the cultural industries.

The people who gain most, materially and culturally, from the cosmopolitan experience of Spitalfields remain mostly young white middle-class professionals. And few of them ever truly engage with the other cultures around them. However, whilst the boundaries of class and ethnic identities are reinforced and reproduced in Spitalfields, I still maintain that sexual identities can, at times, become more fluid there. The multiple ways in which different individuals and social groups live out their roles in the area can occasionally escape established social norms and open up possibilities to think differently about how personal identities are performed in particular kinds of social space. In particular, although the presence of gay space is downplayed in official representations of Spitalfields, the resulting array of leisure spaces provides room for a layer of non-heterosexual men, for whom mainstream 'gay' identities have outlived their usefulness, to enjoy a more fluid set of encounters in the local bars. Similarly, at times, the presence of a range of relaxed, mixed venues, in which the consumption of alcohol does not always predominate, has afforded social opportunities for some young non-heterosexual Bengali men in the area. For them, cosmopolitan space offers a degree of camouflage in which to explore and perform queer identities.

References

Ali, M. (2003) *Brick Lane*, London: Doubleday.

Amin, A. (2002) 'Ethnicity and the multicultural city: living with diversity', *Environment and Planning A*, 34: 959–80.

Binnie, J. and Skeggs, B. (2004) 'Cosmopolitan knowledge and the production and consumption of sexualized space: Manchester's gay village', *Sociological Review*, 52: 39–61.

Bishop, B. (2000) 'Technology and tolerance: Austin hallmarks', *Austin American-Statesman*, 25 June.

Cityside Regeneration (2002) *Cityside SRB3 Final Report*, London: Cityside Regeneration Ltd.

Delany, S. (1999) *Times Square Red, Times Square Blue*, New York: New York University Press.

East London Advertiser (2004) 'Two gay men viciously beaten by thugs in unprovoked attack', 20 May, p. 5.

Florida, R. (2002) *The Rise of the Creative Class: And How it's Transforming Work, Leisure, Community and Everyday Life*, New York: Basic Books.

Hannerz, U. (1992) *Cultural Complexity: Studies in the Social Organization of Meaning*, New York: Columbia University Press.

Hennessy, R. (2000) *Profit and Pleasure: Sexual Identities in Late Capitalism*, New York: Routledge.

Jacobs, J. M. (1996) *Edge of Empire: Postcolonialism and the City*, London: Routledge.

Lefebvre, H. (1991) *The Production of Space*, trans. D. Nicholson-Smith, Oxford: Blackwell.

Lichtenstein, R. and Sinclair, I. (1999) *Rodinsky's Room*, London: Granta Books.

Mitchell, K. (2003) 'Educating the national citizen in neoliberal times: from the multicultural self to the strategic cosmopolitan', *Transactions of the Institute of British Geographers*, 28: 387–403.

Pinder, D. (2001) 'Ghostly footsteps: voices, memories and walks in the city', *Ecumene*, 8: 1–19.

Rushbrook, D. (2002) 'Cities, queer space, and the cosmopolitan tourist', *GLQ*, 8: 183–206.

Shoreditch Map Company (2004) *The Shoreditch Map*, 67 (February), London: Shoreditch Map Company.

Simpson, M. (2002) 'The return of the metrosexual', in M. Simpson, *Sex Terror: Erotic Misadventures in Pop Culture*, Binghampton, NY: Harrington Park Press.

Sinfield, A. (1998) *Gay and After*, London: Serpent's Tail.

8 Negotiating cosmopolitanism in Singapore's fictional landscape

Serene Tan and Brenda S. A. Yeoh

Introduction

The notion of cosmopolitanism as an important ingredient in the success of Singapore as a nation in the global arena is an issue of recent and popular interest. The 'Singapore 21 Vision' launched by Prime Minister Goh Chok Tong (henceforth PM Goh) in 1997 envisions Singapore as the ultimate cosmopolitan pit stop. This vision of the city aims to strengthen the 'heartware' of Singapore, drawing on the intangibles of society such as social cohesion, political stability and the collective will, values and attitudes of Singaporeans. This message has been communicated through constant pronouncements that the city should aim to become a 'cosmopolis' (Mah 1999), a 'Renaissance City' (Goh 1999), a 'global city' and a 'globapolis' (Goh 2001). Emphasis has been placed on 'cosmopolitans' who are 'indispensable in generating wealth for Singapore' (Goh 1999). The developing discourse on cosmopolitanism in Singapore is reflected in a number of specific ways. First, while some have agreed that the world is entering a post-nationalism era, it is clear that the strings of nationalism are still clearly attached in Singapore. The government intends to have a cosmopolitan Singapore, yet wants its citizens to feel emotionally attached to and identify with the country. As articulated in the 'Singapore 21 Vision', 'whether [Singaporeans] live in Singapore or overseas . . . [they] must develop stronger bonds of belonging and commitment to [the] country' (Singapore 21 Committee 1999). Alongside cosmopolitan goals, the demands of nation-building are also imperative, if somewhat contradictory.

Often the discourse on cosmopolitanism is contrasted with that of 'heartlander-ism'. This cosmopolitan–heartlander divide sets better-educated, highly mobile 'cosmopolitans' against more rooted, less mobile 'heartlanders'. While both are said to be important to the overall fabric of Singapore, cosmopolitanism is implicitly seen to be the more exalted vision to aim for. While Singaporeans are heartlanders almost by virtue of nature and circumstance, cosmopolitans must be nurtured and developed through being exposed to the world.

The local arts scene provides a circuit of public discourse on the debates on cosmopolitanism. Literature is one of the avenues through which artistes articulate their thoughts and comments. Many have published poetry and plays, but, more popularly, there is fictional prose, through which authors fashion a landscape to reflect local society. This chapter, therefore, explores Singaporean sentiments about the nation's cosmopolitan ideal through the fiction of contemporary Singaporean authors – a group of contributors critical to the shaping of the national imaginary (Singh 1998). The wide range of texts available develops a voice through fiction that comments critically and creatively on Singapore society. The first objective of the chapter considers, using Singaporean fiction as a lens, the 'cosmopolising' of the people of Singapore, and their place in the city as opposed to that of the heartlanders. The second objective examines the emergence of landscapes of cosmopolitanism in Singapore and contrasts it with 'heartland' spaces.

Conceptualising Singapore-style cosmopolitanism

Cosmopolitanism, innately related to geography, implies the presence of a *geography*. Hence, there is a non-homogeneous landscape that would-be cosmopolitans must learn to navigate. Geography creates unequal distribution of resources, which leads to an unequal distribution of opportunity, and this in turn results in unequal power relations. Cosmopolitanism, however, attempts to be a solution to this inequality. Some form of geographical knowledge is presumed in every form of cosmopolitanism (Harvey 2000), and almost any use of 'cosmopolitanism' implies some embedded geopolitical allegory (Wilson 1998, in Harvey 2000). Cosmopolitanism has to confront and defeat the prejudices and oppression that arise in the real, inhabited, geographical world. As the production of geographical knowledges must engage a political project, meaningful cosmopolitanism also involves a political project – to confront the 'banality of geographical evils' (Harvey 2000).

Appadurai (1996) recognises that cosmopolitanism necessarily involves deterritorialisation and globalising movements which operate to transcend specific territorial boundaries and identities. He links this to the cultural dynamics of the global ethnoscape of a world that is rapidly losing its concrete boundaries. The increasing fluidity of cultural structures is symbolised by the transnationalities of people who, as true cosmopolitans, have acquired what Hannerz (1996) calls 'cultural competence' – a built-up skill in manoeuvring with a particular system of meanings, or *cultures* – in this context. Indeed, Cartier (1999) locates geography in cosmopolitanism as a necessary process. Geography is located in processes that have created world cities – places belonging to contemporary global imaginations. As such, it is the cosmopolitan who has created opportunities for such world cities to be formed, and for globalisation to occur. Because of the enterprising cosmopolitan, people of diverse classes and cultural identities are able to exist together in the world economy, forming social, political and economic organisations that reflect conditions of high mobility. These reside in cities whose sense of place is global. Hence, cosmopolitanism is the outcome of specific geographical processes, like the movements of people, ideas and money. This results in changes, transforming places, ways of life and world views (Cartier 1999).

To locate cosmopolitanism in Singaporean terms requires some clarification, for Singapore-style cosmopolitanism is located within and serves the social, economic and political purposes of the country. With the advent of globalisation and the rise of technology that heralds the era of hyper-connectivity and global villages, Singapore races to keep up with the times, placing extra emphasis on the economy and productivity among other values like social cohesion and political stability. The constant urge to be a successful nation in all areas is the driving force of this country. Much of PM Goh's National Day Rally Speech in 1999 focused on building a first-world economy and a world-class home. This is with the intent of making Singapore 'an oasis of talent' (Goh 1999) and the key global node in the Asian region.

As such, we posit that there is a two-pronged approach to cosmopolising Singapore. The first is to make Singapore a place for cosmopolitans and the second is to create cosmopolitan Singaporeans. Making

Singapore a place for cosmopolitans means changing the landscape of the country to suit the demands of a cosmopolitan nation. 'The institutional and social infrastructure [of the city must] facilitate and support the Cosmopolitan Singaporean' (Singapore 21 Committee 1999). In the physical sense, cosmopolising Singapore entails cosmopolising the landscape. This has resulted in Singapore's highly developed and extremely urbanised landscape being described as cosmopolitan – as Kwok (2001) mentions, 'The city is the epitome of cosmopolitan modernity.' In this sense, cosmopolitanism in Singapore is popularly seen, by the government and by the press, in terms of globalisation, urbanisation, industrialisation, modernity, high-speed connections and advanced technology, accessibility and efficiency. With these characteristics, the country, in having a magnetic attraction for people and ideas, hopes to be able to attract, retain and absorb talent from all over the world, whether in business, academia or the performing arts.

Hence, new policies are aimed at developing the economy even more, such as promoting the arts scene in an attempt to create a Singaporean 'Renaissance City' (Lee 2000), thus attracting foreign talent not just in the business sectors, but also in the performing arts and cultural sectors. As Chang (2000: 818) observes, 'A relatively new area in Singapore's globalisation thrust is arts, culture, and entertainment.' This is also a bid to boost the tourism economy, to make Singapore not just an economic hub, but also a performance hub that people from the region would flock to, attracted by the diversity of entertainment and cultural enrichment provided (Chang 2000). As such, Singapore is developing its urban areas, creating and designing urban spaces that will attract the cosmopolitan foreign crowd, as well as landscapes that are functional, efficient and attractive. Many examples can be seen in the high-rise office blocks and the world-class shopping malls, but to add to this, a new and spectacular arts and cultural centre, 'The Esplanade – Theatres on the Bay', has been constructed to support the reinvention of Singapore as a 'Renaissance City' (Chang 2000).

Creating cosmopolitan Singaporeans is just as important. 'Singaporeans need . . . to go global', and to be 'world ready, able to plug-and-play with confidence in the global economy' (Singapore 21 Committee 1999). This deals with the social-cultural sector of the nation. Singaporeans must be cosmopolitan in order to feel comfortable with the rest of the world – and this is so that Singaporeans can go into the rest of the world to promote the Singaporean cause of becoming a successful nation. Thus, physical cosmopolitanism is related to cultural cosmopolitanism. As Cartier (1999: 279) notes, 'If the cosmopolitan has a geography, world cities are its primary nodes.' In being the node for the best ideas from around the world, the physical infrastructure of cosmopolitanism is not enough. A vibrancy of culture is required in constructing a cosmopolitan city. In making Singapore a global city where peoples of different nationalities and backgrounds are able to work together in harmony, success is to be ensured by making 'the national spirit of the country a cosmopolitan one' (Yeo 1994).

Cultural cosmopolitanism also means creating an openness to peoples of all nationalities, where the East meets with the West, and where the influences of the great cultures that Singapore has absorbed are arranged in a kaleidoscope of multiculturalism, multiracialism, multiethnicism and multilinguisticism. As Goh (1995) mentions, the 'Chineseness' and Confucianist values are as essential as the 'Malayness', 'Indianness' and Western orientation of Singaporeans in making Singapore a uniquely cosmopolitan nation. The widely promoted harmony in multiracialism in Singapore is based on the premise and condition of the Singaporean's ability to feel comfortable with peoples of differentiated race, cultures and traditions, and more than this – to share and partake with them. With this innate acceptance of others, the ideal Singaporean is thus equipped to be the model cosmopolitan. The cosmopolitan Singaporean is touted as having the ability to think globally while acting and feeling locally. This is suggestive of a compromise between cosmopolitanism and nationalism. The cosmopolitan Singaporean would be familiar with global trends and lifestyles, diverse cultures and social norms, and feel comfortable working and living in Singapore as well as overseas and would grow up with an international perspective. Internationalisation is a key word here – the internationalised Singaporean is a cosmopolitan Singaporean.

The cosmopolitan–heartlander dialogue took distinct shape in PM Goh's National Day Rally Speech in 1999. He mentioned that there exist two types of Singaporeans – the cosmopolitan and the heartlander. He defines the cosmopolitan, as mentioned before, as international in perspective and contributing to the generation of wealth for Singapore. The heartlanders, however, make their living within Singapore and have local orientation and interests rather than international ones. They have skills that are, ironically, globally relevant, but that are perceived as unmarketable beyond the region. In the classification of the state heartlanders speak Singlish (a form of colloquial Singaporean English) – therefore are incomprehensible outside of the country – and include taxi-drivers, stallholders and other blue-collar workers among their ranks. However, PM Goh insists that heartlanders do have their function. They play a major role in maintaining the core values and social stability of the country.

The proponents of humanistic cosmopolitanism envisage it as an alternative to nationalism. As words like *globalisation* and *transnationalism* gain popularity, *nationalism* seems to be rapidly going out of fashion in these circles. Yet the Singaporean ideal of cosmopolitanism seems to include both the cosmopolitan and the nationalist existing together in some kind of mutual existence, combining a global, yet local, outlook in life. In other words, Singapore needs to spread its wings to succeed as a global, cosmopolitan country, while being able to retain its roots and the richness of culture in all its multiracial diversity. The 'Singapore 21 Vision' (Singapore 21 Committee 1999) states that 'the more Singapore becomes internationalised in the 21st century, the more crucial it is to have a strong national heartbeat' and that citizens 'must develop stronger bonds of belonging and commitment to this country'. Perhaps, in this way, the ideal *cosmopolitan Singaporeans* will differ from the *idealistic cosmopolitans* in the sense that they are not a rootless people, but rather a cosmopolitan people with a heart in their country of origin. Cosmopolitan Singaporeans are thus expected to think global and feel local. In this sense, they are then *rooted cosmopolitans* – cosmopolitans who feel comfortable and able to belong in any country, but who still feel rooted, and belong to their place of birth.

A word on the methodology

We approach this investigation of cosmopolitanism through the textual analysis of local Singaporean fiction. Literature provides a valuable way of examining relevant social issues of the day and local Singaporean fiction holds many examples of social commentary. Fiction develops an understanding of a nation's history, and contributes to the shaping of mindsets, attitudes and imaginations (Singh 1998). The author is able to develop insights, approaches and creative spaces through which a story can be told to broaden the awareness of the reader. The approach of this paper gives weight to individual experiences and accounts, and creatively dwells on meanings, turning empty spaces into places filled with significance and value. It is reality coloured by the lenses of experience and influence. Thus fictional texts provide a creative lens with many degrees of freedom through which the sentiments of the people toward cosmopolitanism may be discerned.

The source of primary data for this chapter is drawn from the interpretation of imaginative novels and short stories written with an understanding of the Singapore landscape. Our choice of texts includes novels and short stories published between 1995 and 2002, and written by Singaporean writers. A variety of authors, both young and mature, and genres are selected to give a broad context in which cosmopolitanism can be viewed, and to prevent a bias that could take place as a consequence of having too small a range. For a text to be relevant, it must include dilemmas, like seeking a sense of belonging and deliberations of culture and traditions, which frequently haunt the contemporary Singaporean. Literary devices that are examined in interpreting these works include the author's style of writing, any special literary devices that work to bring across specific messages, and metaphors, tropes and images that regularly arise to connote a theme, or a premise for argument.

Tuan (1978) explains the value of literature to geographical research in three main ways – literature as material, as a model of writing, and as a mode of perspective to experience and culture. As Kong (1986) also mentions, the 'intimate' subjective dimension of human–environment relationships can be revealed more intensely in literature than in social surveys. Pocock (1978) posits that imaginative literature contributes to environmental knowing, being an important ingredient in our anticipation of, and encounter with, places. Writings give rise to 'valuable' landscapes – one which is valued because of associational qualities. Thus the contribution of literature to geography is in the making of places – landscapes laden with meaning.

Textual analysis provides a humanistic approach which offers insights into the cultural landscapes of places. To Barnes and Duncan (1992), 'writing mirrors the world' and is constitutive of reality. They quote Eagleton (1983, in Barnes and Duncan 1992: 2): 'in the ideology of realism . . . words are felt to link up with their thoughts or objects in essentially right and incontrovertible ways: the word becomes the only proper way of viewing this object or experiencing this thought'. As such, literature is a 'reconstruction of experience' (Tuan 1978), which converts space into meaningful place and is also a mode of social commentary.

Singaporean literature in English has been a largely ignored group of writing in its own country, mainly because there is an innate compulsion to compare these local works with those from England and America. However, the native literature of a country is exceedingly important, because much of its history would be encapsulated in the literature (Singh 1998). In Singapore, there are signs that a very solid base of viable, vibrant literature in English in all the main genres is being built (Singh 1998). Writers write about experiences that are personal and exceedingly distinctive, particularly in a culture like that of Singapore. Literature also plays a part in contributing to social commentary, as most of the texts deal with issues that arise and are highly relevant in Singapore society, and which affect its citizens – the prolific Catherine Lim, among others, constantly writes on the nature of mankind, and the values, conflicts and issues of concern to Singaporeans (Smith 1998).

Cosmopolitan Singaporeans

Through fictional lenses, we consider the place of the cosmopolitans and the heartlanders in the nation, the degree to which the cosmopolitans and the heartlanders are divided, the level of comfort at the zones of contact between them, and the issues that arise out of the division via oppositional themes and contexts.

Rootedness and rootlessness

Edward Relph (1976) defines rootedness as the communal and personal experience of places that produces emotions of close attachment and familiarity to that place. This particular attachment constitutes peoples' roots in places, and the familiarity that this involves does not include just detailed knowledge, but also a sense of deep care and concern for the place. Only through a multiplicity of experiences and interactions with a particular place does a person develop a sense of rootedness to that place. Thus, rootlessness, as its polar opposite, indicates the lack of close attachments, responsibility and respect to a place and what it represents. Heartlanders thus are considered 'rooted', while cosmopolitans, in contrast, remain 'rootless'.

In many of the texts, the main protagonists seem to find themselves rootless. In *Heartland* (Shiau 1999), the protagonist Wing is a heartlander who feels rootless in the midst of the heartlands in

Singapore. Shiau's device of linking the present, Wing's story, to the past, the founding legend of Sang Nila Utama (henceforth, Utama), seems to indicate the condition of rootedness of the Singaporean community to their land. Utama was the Palemban prince who was the first Malay king of Singapore, and who also gave the country its name, according to the Malay Annals. The story illustrates the disjuncture between the prince who found the land and ruled it and the boy who was born in the land yet knows not if his place is truly there. The Singaporean community can be considered immigrants because the island, previously known as Temasek, was uninhabited when Utama founded it. At times, the text suggests that the citizens of Singapore *are* rooted to the country – 'the immigrants set their roots in the heartlands' (Shiau 1999: 190). Yet the rooting of the immigrants in the heartlands is complicated and problematic, for, at the same time, the author throws doubts on the plausibility of these immigrants belonging to this country. For a moment, Wing experiences doubt with regard to his place in the country when he finds that the man his mother was married to was not his father, because, lacking an origin, he did not know where he belonged. If Utama 'had created an illegitimate heirloom' (Shiau 1999: 188), then there is the possibility of the condition of rootlessness that affects all heartlanders, as their claim on the country would have then been misplaced.

Shiau also presents a contrasting view of the rooted and the rootless, like Utama, who made the land his home, and Wing, who is not sure where his home is. While May Ling and her family are presented as the extreme heartlanders, the people who are clearly rooted in the land and have no doubts about their identity, and Chloe and her family, the rootless, are from the other extreme end occupied by the cosmopolitans, others are less comfortable with their identities and their places in the world.

In contrast, the protagonist Deng in *Mammon, Inc.* (Tan 2001) is clearly rootless and cosmopolitan – she does not quite know where she belongs, and has a very limited sense of identity. Brought up in Singapore, educated in Oxford, and offered a job in New York with Mammon, Inc., the largest corporate entity in the world, Deng is caught between her Eastern/Chinese heritage and her Western/Christian education and way of life. The corporate emblem of Mammon, Inc. is a dragon, which in the East signifies wealth and prosperity, but in the West represents evil. While Singapore has its material comforts – like the food that she enjoys – she is ultimately uncomfortable here. Instead, she has progressed culturally to adapt to the different places she lives in and the different people she meets. She is also slightly ill at ease in the great corporation, Mammon, Inc., which is based in New York. The position that Deng is offered is that of an 'adapter': someone who adapts people to cultures other than their own. These people, called *global nomads*, are the 'modern international elite' (Tan 2001: 2): executives who grew up in one country, were educated in another, and currently work in a third. When these *nomads* relocate to another country, they need help learning how to gain social acceptance, and an adapter will be assigned to them to help them adapt. Eventually Deng chooses to work with the mammoth corporation rather than remaining with her Christian professor at Oxford, and as such stakes a claim on a cosmopolitan identity, because that is exactly what the mammoth corporation offers. Mammon, Inc. offers an entirely cosmopolitan way of life. Deng's vision of what she could be if she joined the corporation is that of 'a cosmopolitan jet setter, equally at home with chiefs in mud huts or tycoons in the Four Seasons' (Tan 2001: 71).

Conversely, in *Foreign Bodies* (Tan 1997), a more complicated question of identities arises. Mei feels slightly rootless and uncomfortable in Singapore with her high-paying job in Raffles Place, while Andy, a British man, manages to make himself feel comfortable by integrating with the community. As such, Mei feels like a foreign body in her own country, whereas Andy becomes almost local. Thus, Mei is the cosmopolitan who seems truly rootless, unable to settle anywhere, yet Andy is a cosmopolitan of a different sort – able to feel comfortable everywhere, able to belong. However, at the end of the novel, the roles are reversed. Mei once again slips back into her Singaporean identity, doing things the Singapore way by refusing to become deeply involved in other peoples' business. Understanding the ways of the Singaporean court of law, knowing she cannot fight the system, she conforms. Andy is finally ostracised by the Singaporean community that distrusts foreigners and which would not consider the truthful word of a 'fallen' stranger, but would rather rely on its local golden child – the well-educated, high-class elite of the country no matter how dubious they appear. Within this text, it is realised that there are people who are cosmopolitan in nature, but who still feel rooted to the country just as Mei feels. This indicates that different degrees of cosmopolitanism exist – the cosmopolitan who fits in anywhere, and the cosmopolitan who does not. There are very rarely any cases of absolute cosmopolitans or pure heartlanders.

The condition of rootedness and rootlessness can be seen in both the cosmopolitans and the heartlanders. The heartlanders, rooted to the land, seem secure in their place; as do the cosmopolitans who are comfortable with being rootless. But there also exist those who do not fall comfortably into such well-defined dichotomies. There are confused heartlanders who feel rooted, but are also aimless and lacking an identification with the land; likewise, there are cosmopolitans who are mobile, and are able to fit in comfortably anywhere in the world, yet also feel attached to certain places. This complexity troubles the government's dualistic divide of cosmopolitans and heartlanders. The lack of openness of rooted heartlanders to foreigners depicts Singaporeans who are unable to adapt to globalisation. Heartlanders who experience doubts about their roots stay in the country but cannot identify with its dreams. The flight of the rootless cosmopolitans from the land indicates a culture of disassociation. Yet there is hope while the rooted cosmopolitan, epitome of hybrid fluidity, holds true to his or her attachment to Singapore, the prime example of the ideal Singaporean.

Contact zones

Contact zones indicate social spaces where different cultures meet and interact with each other, and, as Pratt (1992) states, often in highly asymmetrical relations of domination and subordination – exemplified in colonialism, slavery or their aftermaths as they are lived out across the globe today. Pratt (1992) also defines contact zones and the space in which geographically and historically separated peoples come into contact with each other and establish on-going relations. It is an attempt to recreate the spatial and temporal presence of subjects previously discrete, but whose existences now meet and intersect. The use of the term contact zone in this chapter highlights the interaction forced by contact between the peoples commonly classified as cosmopolitans and the peoples typically stereotyped as heartlanders. The 'contact' perspective emphasises how the cosmopolitans and heartlanders are seen in the context of and by their relations to each other.

Many of the texts demonstrate friction at the boundaries and discomfort between the cosmopolitans and the heartlanders. The evident disparity between the two categories of people is raised in issues and themes, like the inability to communicate easily on the same level; the different needs, wants, expectations and ambitions of the people; and the different spaces in which they exist.

The inability to communicate is evident in many of the texts. In *Heartland*, Wing's inability to communicate with his first girlfriend, Chloe, shows this. Chloe is cosmopolitan – she sees her future outside of Singapore, and in many places the texts hints at social class differences between Chloe and Wing. Though the relationship started easily, the couple soon run out of things to say. Furthermore, Chloe is unable to communicate with Wing's mother, who speaks only in dialect, whereas the heartlander May Ling, Wing's later girlfriend, communicates easily with her. For this reason a rift is created between Wing and Chloe. Different needs, wants, expectations and ambitions are reflected in *Heartland*, often in contrast. Wing's girlfriend Chloe and her family are unable to accept him because he seems to have neither ambition nor direction for his future; in contrast, Chloe already has plans to take her education further, to leave Singapore and study overseas. Chloe and her family represent the cosmopolitan, elite family – her father is a successful businessman who contributes much, financially, to the country. She is the polar opposite of May Ling, who is simple, not well educated, and who works as a shop attendant. May Ling has few needs and wants, is not as ambitious, and expects less of Wing as compared to Chloe. Even their names are significant. May Ling has a Chinese name, and she has roots in Singapore; Chloe has an English name with French origins, and she is cosmopolitan, like her name.

In *Skimming* (Tham 1999), David, who feels cosmopolitan, is unable to communicate with his heartlander family, and he feels estranged from them. He also hides his heartlander background, choosing not to reveal them to the cosmopolitan Li and Keong, and this signals a break in the open, sharing communication that the friends have. In both *Mammon, Inc.* and *Foreign Bodies*, the main protagonists of the stories are sometimes unable to communicate entirely with the people around them. Deng in *Mammon, Inc.* is sometimes misunderstood by both her best friend, the British Steve, and her heartland Singaporean sister, Chen. Steve cannot understand her *kiasu* nature (colloquial Singaporean English meaning 'afraid to lose', which is seen to be a typical Singaporean trait), and Chen is ignorant of the reasons behind her indecisiveness in choosing between Mammon, Inc. and Professor Ad-Oy. Meanwhile, Mei from *Foreign Bodies* has problems communicating with her mother, who, in turn, has problems understanding her, the 'strange things' that she does, and why she does them.

In *Following the Wrong God Home* (Lim 2001), the discomfort between different people is also clearly registered. Yin Ling has an affair with the foreigner Ben, because she can communicate and connect with him, a cosmopolitan with whom certain heartlanders are uncomfortable. Ironically, she is attracted to him because he is able to fulfil her Singaporean and culturally rooted needs, like taking care of her aging nanny, Ah Heng Cheh, and her god, better than Vincent, her Singaporean husband, no matter how much care the latter lavishes on her, ever did. Yin Ling struggles to reconcile her cosmopolitan yearnings, to be with Ben and to live his kind of lifestyle, with her traditional cultural roots, to stay devoted and loyal to the family that she has married into. In a strange reversal of themes, the foreign cosmopolitans symbolised by Ben and his foreign friends, like the Jamiesons who have conformed to the Singaporean lifestyle, are more comfortable with the traditional heartlander, Ah Heng Cheh. Ben understands the needs of Ah Heng Cheh, and tries to help her fulfil them, whereas other Singaporeans, like Vincent's mother, are unable to accept her, even though they are of the same race and culture.

The theme of the discomfort of cosmopolitans with heartlanders manifests itself in many ways. The inability to communicate and the differing expectations of the two groups seem to make the point that cosmopolitans and heartlanders are unable to mix, that the boundary where the two meet is rough with friction: they are inherently uncomfortable with each other. Yet there are rare instances that contrast to this, like Ben's comfort with Ah Heng Cheh, and Andy's ability to feel comfortable in Singapore. While discomfort is expected between the cosmopolitans and the heartlanders, it is not expected that foreign cosmopolitans should feel, oddly, comfortable with Singaporean heartlanders. This seems to hint that Singaporeans find it harder to reconcile the fact that their country is divided, and are unable to bridge the gap, while foreigners are strangely more accommodating. A complete foreigner and a completely traditional heartlander, who are secure in their respective identities – unlike Singaporeans like Wing and Mei who are struggling with their place in Singapore and their identity – seem to be less confused and more unable to cope with difference.

The complexity of emotions and sentiments that cosmopolitan Singaporeans, as shown in the texts, have displayed is an indication of how they can be rooted, yet rootless at the same time. Their rootlessness is an indication of cosmopolitan yearnings, of their mobility and the ability to fit in with other societies in other countries and nationalities; yet their rootedness is an indication of the home that they belong to – and this is something that most of them, although in conflict with it, are unable to let go. This sentiment is echoed in *Heartland*, where Wing feels that '[Singapore's ancestors] were like migrants whose pride is borrowed rather than historical. And like migrants, they carved a little niche for themselves like the Chinatowns and Little Italys that spring up all over the world' (Shiau 1999: 12). People feel a need to belong, a desire to have roots, a place to call home.

On the conflicts at home, the struggles that characters like Deng and Mei have with the system and the culture in their home countries are manifested in their discomfort and their inability to fit in completely once they have become more cosmopolitan than their peers. Thus the degree of cosmopolitanism, relative to others around them, that each individual possesses seems to have the ultimate say in how well they fit in. While Deng is able to straddle the divide between inside and outside – her family in Singapore and her foreign friends, notably Steve – she is unable to reconcile the two. No matter how hard Deng tries, at the very basic level, her parents reject Steve when he joins her family for dinner at a hawker centre. The gulf of discomfort exhibited by Deng's parents, compared with the easy friendship that develops between Andy and Deng's sister Chen, shows that there are Singaporeans who resist the cosmopolitan ideals of the country. Thus the inability of foreigners to break into the Singaporean way of living – also showcased in *Foreign Bodies* and *Following the Wrong God Home* – suggests that Singapore may not be such a cosmopolitan country after all. The lack of a general vote of acceptance of foreigners into the country indicates a certain hostility that prevents people like Andy, Ben and Steve from fitting completely into Singaporean society, customs and traditions. Rather, these defeat them.

At times, Singapore seems to be a sort of undeveloped, backward-oriented country, where most of the citizens are unable to communicate with outsiders and foreigners – in *Mammon, Inc.* Chen struggles in vain to fit in with the British elite. It can be said that this is the stereotypical view of heartlander-Singaporeans – that they are simply unable to be cosmopolitan, flexible and adaptable with the rest of the world. Yet this may be an indicator of how the cosmopolitan Singaporean views Singapore – as an island of people who are categorised into groups. Thus the cosmopolitan and the heartlander divide people who are judged by their (in)ability to conform to the rest of the world.

Cosmopolitan places?

If Singapore is to be a place for cosmopolitans, then the physical landscape of the nation must complement the cosmopolitan natures of its citizens. This section discovers, through the lenses of the authors, the spaces of Singapore which are, and are not, cosmopolitan, and how exactly these spaces are conceived. The degree to which Singapore can be considered cosmopolitan also emerges through narratives about its landscapes. The texts reveal both clearly defined spaces, such as the city areas which are considered cosmopolitan space, and the other places, such as the heartlands, or the suburban areas categorised as heartland spaces.

Cosmopolitan space

At the global level, Singapore is modern, advanced, and the technological equivalent to many other developed countries. The landscape of Singapore is described as 'a sight of superlatives'; it has 'the tallest hotel in the world . . . in South East Asia, a white shard among the silver skyscrapers' (Shiau 1999: 212). Singapore's modernity is always described in terms of its infrastructure, so in terms of landscape, on a visible whole, Singapore is definitely cosmopolitan on the global scale.

On the same level, the internal debates about its cosmopolitanism can be seen in *Foreign Bodies*:

> A few hundred metres from the bridge, I could see two men in long red robes fighting on a stage. They were doing 'wayang' . . . From my office, I could see the Westin Stamford, the tallest hotel in the world. The other skyscrapers towered above the green waters, white light bouncing off dark windows, a labyrinth of mirrored citadels, a city of glass. This was Singapore, the centre of information technology in South-East Asia; this was Singapore, a place where people still bowed down to idols, burnt joss sticks, consulted mediums, exorcised demons and walked on coals.
>
> (Tan 1997: 147–8)

This illustrates the universal description of Singapore, well known as a melting pot of cultures. On this scale, it is clear that traditional practices in Singapore – the images of 'wayang', of mediums, and other religious practices – coexist with its global image as a place of advanced infrastructure and information technology. This comfortable coexistence is the very picture of cosmopolitanism.

On the local level, Singapore is divided up into little colonies of cosmopolitan areas and heartland pockets. As stated earlier, the landscape is, at the most basic level, divided into a city-cosmopolitan/suburb-heartland dichotomy. Clearly defined descriptions of cosmopolitan spaces are obvious in some of the texts, such as the Central Business District, where Mei works, in *Foreign Bodies*, or Orchard Road and its several shopping centres, like The Heeren, in Sa'at's short story, *Cubicle*, and in Tan's *Mammon, Inc.* A prime example is Bugis Junction and the nearby hotel in Shiau's *Heartland*, where Wing loiters in the opening of the novel:

> The door to the Hotel Inter-Continental swung open . . . occupied by only two idle tai-tais with Louis Feraud paper bags and some Caucasian men in suits . . . the subtle oriental touches to the European décor, together with the piped-in Bach, gave the place an air of colonialism.
>
> (Shiau 1999: 4)

This shopping centre and hotel are clearly cosmopolitan; the expected mix of cultures and nationalities – local tai-tais, French shopping, Caucasian men and oriental touches on the European décor, that together occupy a single space without chaos – are indicative of a cosmopolitan ethos. The lack of friction between the cultures that are represented here gives rise to an air of comfortable acceptance that the cosmopolitan sentiment is known for.

This is greatly contrasted with the heartlands that Wing returns to later:

> the majesty of the four point blocks piercing the orange dusk sky and the lower blocks, old but graceful, rising proudly from the soil . . . the [MRT station] turnstiles opening and closing for thousands of heartlanders spilling out into the estates . . . many gathered around the old newspaper vendor and her makeshift mahjong-table stall for evening papers . . . the symphony of iron gates . . . TVs . . . fish thrown into woks, metallic quivering of aluminium bathroom doors.
>
> (Shiau 1999: 5–6)

These are 'the sounds . . . like a pulse of life . . . the heartbeat of the estate' (Shiau 1999: 6) that Wing returns to. The heartlands, the HDB homes of the nation, are not quiet, monotonous places, but places where there is life, places to which the sounds and activity of the city return when the sky turns dark. These are places of common domesticity, where the basic needs of the 'natives' are met. The heartlands are the solace of the people; they have nothing to do with the cosmopolitan cares and concerns of the nation. They simply exist for the populace, to serve them, and to provide them with a home.

Heartland place

Once again, at the local level, the texts demonstrate the divide between cosmopolitan and heartland spaces. On the global scale the country is an amalgamation that is most easily described as cosmopolitan, but on the local scale the heartland spaces appear as clearly discrete. It is only at this local level that the disparity between cosmopolitan and heartland spaces becomes clear, and these are the spaces that the characters in the texts experience as they traverse the stories.

The people who live in the present fictional landscape feel the shift in its character when they move from one space to another, from a cosmopolitan to a heartland place. Like Wing, who travels from the cosmopolitan Bugis Junction and the Hotel Inter-Continental to Ghim Moh, one of the oldest housing estates in the country, they have their roots in the heartland spaces, but their expectations are ingrained in the future – the future of the country, in modernity, in the city. In a way, the citizens of Singapore are a reflection of Singapore itself – a country that is mixed, with migrant roots. The country is made up of people who find their own meaning in an intense attachment to their surrounding landscape – and if the landscape is varied, then so is the population. Andy in *Foreign Bodies*, when he first arrives in Singapore, believes that he has found 'Shangri-La' (Tan 1997: 213). It is 'hot and bright', and 'everyone and everything is open', but when Andy settles in, becomes part of the landscape, and tries to fit in, 'Shangri-La has turned to Shit Hotel' (Tan 1997: 214). The vastly varied responses that one acquires from the landscape are remarkable in their diversity.

The past contains the roots of the heartlanders, and there are places still caught in this time frame. The heartlands are characterised by a landscape of public-sector high-rise Housing Development Board (HDB) estates, filled with a daily routine of domesticity. These can be seen in Wing's description of Ghim Moh, his home, and Sa'at's (1999) depiction of the estates in *Corridor*. In *Following the Wrong God Home*, the route taken to the church for the wedding passes by 'one of the oldest housing estates in Singapore' (Lim 2001: 8). It is described as derelict, desolate and old, and Yin Ling connects the landscape to the 'old and retired or young and embittered' (Lim 2001: 8) inhabitants of the estate. The atmosphere that arises from the description is one of squalid dismay, of abandonment, of shame – something to be hidden from foreign eyes, and a sight that conflicts with the nation's cosmopolitanism. Places like the older-generation HDB housing estates, clustered together like buildings seeking shelter from the storm, the hawker centres and the coffee shops which serve a mix of Chinese, Indian and Malay food, where the locals speak in all sorts of dialects – the society that lives in the land contributes to its heartland flavour. As revealed in *Heartland*, through the juxtaposition of the legendary myths of the founding of Singapore, a large part of Singapore is founded in its historical roots. The places where the founders set up their *istana*, or their seats of power – like Sang Nila Utama did on 'Bukit Larangan, later to be called Fort Canning Hill' (Shiau 1999: 77) – were strongholds that protected the populace as long as the ruler stayed true to the land. Thus the past contains the roots of the people, and while these are strong and true to the land, they will keep the nation alive.

Depictions of heartland spaces are unquestionably more ubiquitous than those of cosmopolitan spaces. Entire texts, like *Corridor*, have been devoted to the topic of heartland spaces. In all of Sa'at's stories the images stem from the HDB living spaces, and here, even in the title, Sa'at confines the HDB landscapes to the heartlands of Singapore. In the text, the majority of the protagonists reside in the heartlander HDB flats. His depiction of the HDB lifestyle is almost completely *uncosmopolitan*. There are few, if any, references to foreigners, strangers, a cosmopolitan lifestyle, or cosmopolitan Singaporeans in any way. The stories are based solely on the private lives of the people living in these flats, who have no contact with, or even a thought for, the outside world at large.

HDB flats and their surrounding areas are often taken to be synonymous with heartland spaces. Nowhere is there a description of the HDB flats as other than what they are – dwellings of native Singaporeans, most of whom have little or no involvement with cosmopolitanism of any kind, in any way. *Duel*, one of Sa'at's short stories, is centred upon an unnamed protagonist's fascination in the small hours of the night with a lighted window in the HDB block opposite to his. The detailed description of the corridors connecting the HDB flats in the short story *Corridor* highlights the monotony of life in the heartlands. And it is not just the HDB blocks and the flats, but also the hawker centres and the coffee shops in these areas. The coffee shop that Wing frequents is undeniably heartland. It is old, patronised by the rest of the neighbourhood – by people who speak dialect, by *heartlanders*, by people who hold simple jobs. In *Foreign Bodies*, the hawker centre to which Deng brings Steve, to meet her parents and to experience a slice of 'true' Singapore, is also a heartland space. As Deng tells Steve after he complains about the atmosphere, 'I wanted to bring you to some place that wasn't so clean and sterile . . . break the Singapore stereotype', and she also explains that she 'liked the riotous untidiness of the hawker centres', and describes it as a place where 'you chucked all protocol down the drain outside' (Tan 1997: 221), and so forth. The extreme heartland characteristics of places like these – that cater exclusively to the heartlander population – allow for a true experience of what it is like to be Singaporean: as Steve realises, not simply through the atmosphere and his surroundings, but also through the food.

The different spaces which the heartlanders and the cosmopolitans occupy are also indicators of the disparity between the two. The heartlands are a representation of the past, while the cosmopolitan areas – the high-rise buildings in the CBD – are a representation of the country's present and future. Heartlanders mainly live in the *heartlands* of Singapore, in the ubiquitous HDB blocks that dot much of the country, while most cosmopolitans live in the private condominiums and landed properties in the prestigious and expensive districts that are close to the urban centres. Wai Keong and Li in *Skimming*, both from distinguished cosmopolitan families, live in private apartments and landed properties, while David, being cosmopolitan, rejects the HDB flat that his family lives in. *Heartland*'s Wing and May Ling both live in Ghim Moh, an old estate in the heartlands, and they are comfortable with each other. Chloe tries to deny that class difference (often indicated by where one lives) will affect their relationship, but becomes excited when she thinks that Wing lives in Mount Sinai, a prestigious part of the country; however, when he corrects her, she begins to feel mildly uncomfortable. She is used to eating in expensive places, while Wing tries to keep their dates affordable. The relationship breaks down finally at an old coffee shop in which Wing feels completely comfortable, while Chloe is distinctly disgusted at her surroundings. Further indications of this can be seen when she twists her ankle due to a hole in the HDB parking lot, demonstrating how out of place she is in the heartland environment.

This illustration appears to suggest that the landscape makes the person. If one is from the heartlands, one's fate seems to be that of a heartlander, and vice versa. However, Yin Ling in *Following the*

Wrong God Home disproves this theory. Growing up in a heartland environment, she nevertheless gains a cosmopolitan education, marries into cosmopolitan opulence, and has an affair with a cosmopolitan foreign man. What this suggests is that there may exist within modern Singaporeans an innate desire to be cosmopolitan. This would then imply that the citizens, particularly the younger ones, feel a need to conform to Singapore's rapidly evolving landscape. If one identifies with his or her surroundings to form his or her identity, then one would naturally take on the characteristics of the landscape.

Conclusion

This chapter has examined the creative ways in which a select group of voices – local Singaporean authors – have chosen to depict the concept of the nation's cosmopolitan ideal. The effects of globalisation on a geographical scale have led to the cosmopolising of Singapore, not simply as the result of natural progression but also through the visions, plans and policies of the government. The nation's visions of becoming an ideal cosmopolitan nation have been held up against the reality that is recognised and revealed in the local literature.

As the texts have shown, the condition of rootedness and rootlessness and the nature of the conflicts at the contact zone between the cosmopolitans and the heartlanders indicate that the cosmopolitan–heartlander dialogue is an issue of some concern. The rather confused majority of the population (as symbolised by the characters in the texts), who are divided into broad categories of rooted heartlanders, rootless heartlanders, rootless cosmopolitans and rooted cosmopolitans, implies that the identities of many are not as settled and concrete as the government would like them to be. At the same time, there have arisen Singaporeans who fulfil the nation's cosmopolitan dreams, and these are the rooted cosmopolitans – the people who are able to feel comfortable anywhere in the world, yet retain an emotional connection to Singapore. These are people who would choose to make Singapore their home, in spite of the multitude of alternatives and choices that are open to them.

The emphasis in the texts on the fusion of both cosmopolitan and heartland cultures and characteristics on the Singaporean landscape is also an indicator of the extent to which cosmopolitanism has taken hold of the nation-state. While it seems that the nation has succeeded in making a cosmopolitan landscape, the cosmopolising of the landscape is fragmented into places that have been almost completely transformed by cosmopolitanism and places which have not been touched by its influence. There are also spaces which are still in the process of transformation, from heartland to cosmopolitan, and there are places which are being kept as heartland and prevented from becoming cosmopolitan, to serve as reminders of the nation's past – its heritage.

The linking of people to their place, their landscape, is an important theme. The comfort level of Singapore society to cosmopolitanism is still dependent on the degree to which they feel at ease in a highly mobile landscape, one that is constantly morphing according to the whims and designs of the nation. Singapore appears to be in a state of flux, caught in an 'in-between' stage where the signs of cosmopolitanism are clearly displayed all over the nation, yet there remain areas that cling on to the anachronism that is yesterday – the places and people of the heartlands that still flaunt the influences and impacts of deep historical rooted practices, cultures and traditions. With this sort of mixed heritage and history, the cosmopolising of Singapore can work in two ways: first, to make the nation entirely cosmopolitan and rootless; and, second, ideally, to make the nation cosmopolitan but rooted. For the nation-state to retain its peculiar brand of cosmopolitanism, the country must be rooted in a strong identity that ties them to their place of origin. This is in accordance with Clifford's belief that no-one 'is permanently fixed by his or her "identity"; but neither can one shed specific structures of race and culture, class and caste, gender and sexuality, environment and history' (Clifford 1997: 12).

References

Appadurai, A. (1996) *Modernity at Large: Cultural Dimensions of Globalization*, Minneapolis, MN: University of Minnesota Press.

Barnes, T. J. and Duncan, J. S. (eds) (1992) *Writing Worlds: Discourse, Texts and Metaphors in the Representation of Landscape*, London: Routledge.

Cartier, C. (1999) 'Cosmopolitics and the maritime world city', *Geographical Review*, 89: 278–89.

Chang, T. C. (2000) 'Renaissance revisited: Singapore as a "global city for the arts"', *International Journal of Urban and Regional Research*, 24: 818–31.

Clifford, J. (1997) *Routes: Travel and Translation in the Late Twentieth Century*, Cambridge, MA: Harvard University Press.

Goh, C. T. (1995) Speech at Nantah Global Reunion Dinner, Nanyang Technological University, 3 June.

Goh, C. T. (1999) *Prime Minister's National Day Rally Speech, 1999: First-World Economy, World Class Home*, Singapore: Ministry of Information and the Arts.

Goh, C. T. (2001) *Prime Minister's National Day Rally Speech, 2001: New Singapore*, Singapore: Ministry of Information and the Arts.

Hannerz, U. (1996) *Transnational Connections: Culture, People, Places*, London and New York: Routledge.

Harvey, D. (2000) 'Cosmopolitanism and the banality of geographic evils', *Public Culture*, 12: 529–64.

Kong, L. (1986) 'Environmental cognition: the Malay world in colonial fiction', unpublished manuscript, Department of Geography, Faculty of Arts and Social Sciences, National University of Singapore.

Kwok, K. W. (2001) 'The worlds of war', in *Straits Times*, Singapore, 31 December, p. L5.

Lee, Y. S. (2000) 'Singapore, the Renaissance City', speech in Parliament, 9 March.

Lim, C. (2001) *Following the Wrong God Home*, London: Orion.

Mah, B. T. (1999) *Making Singapore the Best Place to Live, Work and Play*, speech at the official opening of the URA Centre, launch of the EDA system and the book *Home. Work. Play*. Available at www.ura.gov.sg/pr/tables/pr99-29a.html.

Pocock, D. (1978) *Images of the Urban Environment*, London: Macmillan.

Pratt, M. L. (1992) *Imperial Eyes: Travel Writing and Transculturation*, London: Routledge.

Relph, E. (1976) *Place and Placelessness*, London: Pion.

Sa'at, A. (1999) *Corridor: 12 Short Stories*, Singapore: SNP Editions.

Shiau, D. (1999) *Heartland*, Singapore: SNP Editions.

Singapore 21 Committee (1999) *Singapore 21: Together We Make the Difference*, Singapore: Singapore 21 Committee.

Singh, K. (ed.) (1998) *Interlogue: Studies in Singapore Literature*, Singapore: Ethos Books.

Smith, P. (1998) 'The politics of feminist expression: an essay on Catherine Lim's *The Teardrop Story Woman*', in K. Singh (ed.), *Interlogue: Studies in Singapore Literature*, Singapore: Ethos Books.

Tan, H. H. (1997) *Foreign Bodies*, London: Michael Joseph.

Tan, H. H. (2001) *Mammon, Inc.*, London: Michael Joseph.

Tham, C. (1999) *Skimming*, Singapore: Times Books International.

Tuan, Y. F. (1978) *Space and Place: The Perspective of Experience*, Minneapolis, MN: University of Minnesota Press.

Yeo, G. B. Y. (1994) *Singapore Arts Centre: Taking Shape*, speech at opening of the Arts Centre Design Exhibition, 21 July.

PART III

Producing the cosmopolitan city: cultural policy and intervention

Multicultural urban space and the cosmopolitan 'Other': the contested revitalization of Amsterdam's Bijlmermeer

Annemarie Bodaar

Introduction

Cities today are characterized by increasing ethnic and cultural diversity and are said to be multicultural (Sandercock 1998, 2003; Rogers 2000). This raises new questions about how to maintain social cohesion in multicultural cities. New groups in the city challenge previous spatial practices, and identifications with city and nation, thereby creating discomfort and unease among other groups of urban residents. Central to debates in multicultural cities, therefore, are questions about what aspects of difference are considered acceptable and who gets to decide what forms of difference become legitimate. In the name of multiculturalism, local and national state agencies have taken a role in regulating difference by introducing and legitimating various policies and other forms of intervention aimed at managing difference in cities. The multicultural city has become a frame for politics around questions of difference, neighbourhood identity, social inclusion and urban (re)development. This chapter explores some of these issues in relation to what is often seen as a 'non-cosmopolitan', but multiethnic, neighbourhood in Amsterdam: the Bijlmermeer.

In the literature multiculturalism and cosmopolitanism are often conceptualized in opposite terms (e.g. Beck 2002), where multiculturalism is associated with adherence to the culture of the group and cosmopolitanism with individualism. Cosmopolitan ties are loose and multiple, enabling the cosmopolitan to participate in many worlds (Caglar 2002) rather than the more fixed boundaries of the community associated with multiculturalism. In this chapter, however, multiculturalism is understood as a politics, a set of discursive strategies implemented by the state in order to manage increasing difference, and contested on the ground by society when implemented in neighbourhoods. Cosmopolitanism can then be seen as a government and business strategy to market multicultural neighbourhoods. However, it can also be lived by and promoted by residents of multicultural neighbourhoods in their everyday lives both locally and through their connectedness to transnational networks. Therefore, rather than one cosmopolitanism there are coexisting multiple cosmopolitanisms. What is important, then, is which cosmopolitanism becomes dominant and where, and which others are excluded. This chapter suggests that questions of class and ethnicity are crucial in defining the cosmopolitan and the non-cosmopolitan 'Other', and that cosmopolitanism is a spatial practice. Cosmopolitan urbanism, therefore, refers to the localized struggles over difference contributing to the production of cosmopolitan and non-cosmopolitan spaces.

The concentration of immigrants in disadvantaged neighbourhoods and its relation to the production of cosmopolitan and non-cosmopolitan spaces is one of my key concerns in this chapter. Specifically I wish to explore how a large-scale urban renewal project, aimed at re-imaging an apparently non-cosmopolitan space, is at the centre of struggles over difference. The chapter is organized into four sections. The first explores some of the different conceptualizations of cosmopolitanism, exploring the role of race and class – and understandings of the cosmopolitan 'Other' – in these conceptualizations. The second describes the historical development of social policy in the Netherlands aimed at managing difference and its implementation in the production of (non)cosmopolitan space in Amsterdam's Bijlmermeer. The third considers how difference is negotiated within this neighbourhood space, particularly the different approaches to cosmopolitanism adopted by state actors, business interests and residents. Finally, the conclusion discusses the implication of these socio-spatial negotiations of difference for the Bijlmermeer and multicultural cities more generally. The chapter suggests that the decision over what is considered cosmopolitan space is embedded in power relations, allowing for certain forms and expressions of difference, but masking and marginalizing others.

Ethnicity, class and cosmopolitan 'otherness'

Recently there has been a revival in studies of cosmopolitanism (e.g. Harvey 2000; Vertovec and Cohen 2002). Processes of globalization, migration and the condition of multiculturalism have contributed to its return. This chapter addresses some critical issues raised in this literature on cosmopolitanism, rather than providing an exhaustive overview, which will then be elaborated upon drawing on the case study of the Bijlmermeer. Specifically, it will address some of these issues in relation to ethnicity and class.

In the expanding literature on cosmopolitanism the concept has come to mean different things to different people. One definition is of cosmopolitanism as a new form of global democratization. Here cosmopolitanism refers to a new form of citizenship that is no longer centred in the nation-state. Associated with this definition is the view that cosmopolitanism is 'a stance towards diversity itself . . . an orientation, a willingness to engage with the "Other"' (Hannerz 1990: 239). I would suggest that this is a very idealistic understanding of cosmopolitanism. The idea that people are willing to engage with the 'Other' and to revise their perspectives based on what they learn from other cultures is restricted to only a small group of people, and certainly not to the elite cosmopolitan consumer (the visitors to restaurants, theatres and festivals). This leads to a second definition of cosmopolitanism that plays a more central role in this chapter, where it is defined as a form of consumption. This form of cosmopolitanism, in the form of the commodification of certain aspects of cultures, has proved useful to capitalism in its expansion of new markets. It is often associated with elite Western culture (Vertovec and Cohen 2002). Rather than learning from and truly engaging with difference, cosmopolitan consumption entails the consumption of the non-cosmopolitan 'Other'. Cosmopolitan consumption is associated with cultural capital, with being 'educated and sophisticated'. It is 'the' thing to do, to see, or to participate in. Marketing by businesses and the state, and consumption by elite consumers, play important roles in inscribing difference with value. The place for cosmopolitan consumption, according to Tuan, can be understood negatively as a '"flea market" of gaudy, cheap, imitative wares, and anxious, frantic people striving for the cheap thrill and material advantage' (Tuan 1996: 182). The identity category 'cosmopolitan' is then constructed on the basis of what is considered 'non-cosmopolitan' and consumable, the ethnic 'Other'. The cosmopolitan 'Other' then becomes associated with ethnic, working-class populations who are limited in their power to determine which forms of culture should be domesticated, and who are believed to lack the social and cultural capital to consume and appropriate certain forms of difference.

This idea of cosmopolitanism as elitist and appropriated by capitalism can also be critiqued for ignoring the cosmopolitan nature of the practices of the apparently 'non-cosmopolitan' 'Other' such as migrants and working-class ethnic groups (e.g. Robbins 1998a, 1998b; Featherstone 2002; Vertovec and Cohen 2002). These groups are denied cosmopolitan status, and a role in the production of cosmopolitan spaces, by being constructed as the cosmopolitan 'Other'. As working-class communities these migrant groups are believed to be oriented towards the ethnic community rather than welcoming other cultures (Hannerz 1990). But, as Featherstone (2002: 2) argues, we should move beyond this understanding, as:

> we should not just focus on the cosmopolitan experiences of global elites, or the artists, intellectuals, and tourists of the West, but focus on the working-class cosmopolitan migrants who can be seen as equally able to generate cosmopolitan perceptions.

One example is through transnational networks. Robbins (1998a) calls this 'actually existing cosmopolitanism', recognizing the existence of cosmopolitanism among a wide variety of non-elites, especially immigrants. Drawing on these perspectives, cosmopolitanism can be understood as multiple, and should not be restricted to solely the upper class and Western culture, but also allow 'other' voices to play a role in the production of cosmopolitan spaces.

Finally, cosmopolitanism has been associated with cities (Featherstone 2002) as the places where a mixing of difference occurs. Cities are where people, commodities, ideas and cultures from all over the world meet and interact. Therefore, cosmopolitan urbanism refers to the politics surrounding difference in the production of cosmopolitan and non-cosmopolitan spaces. But within cities it is only certain places that become associated with cosmopolitanism, and these are generally not the spaces of immigrant residence, even though, as Tuan (1996: 187) argues, 'it is in these enclaves [of beleaguered communities], where no ordinary tourists would want to go, that we are most likely to find cultural exotica'. This may not be the exotica that the cosmopolitan consumer is interested in, yet in terms of 'true cosmopolitanism' as an openness towards difference, there is a good chance this is present in the everyday life of the residents of these non-cosmopolitan neighbourhoods.

By branding the Bijlmermeer as cosmopolitan the goal is to broaden the appeal of the area to a wider audience, making the neighbourhood appear less threatening and less dangerous in order to bring in new residents and investment. However, as the case study will suggest, cosmopolitanism understood as a form of consumption, and cosmopolitanism understood as lived or 'actually existing' amongst a largely immigrant population, are bound up in negotiations about the re-imaging of the Bijlmermeer.

Re-imaging non-cosmopolitan space: Amsterdam Bijlmermeer

> The current phase of promoting city identities leaves room for ambiguities. The traditional image does not suffice, yet successive waves of migration raise questions about the cultural unity of cities, which tie in with the issues raised by the presence of cultural minorities that have not been totally assimilated. In cities in the UK, the Netherlands, Germany, Belgium, and France multiculturalism and the cultural differences that can be accepted by a society have become central issues.
>
> (Le Galès 2002: 143)

Underlying the production of cosmopolitan and non-cosmopolitan spaces are decisions about the particular forms of difference that can be accepted. The state, for example in urban policy programmes, plays an important role in regulating ethnic diversity. This section discusses the context of Dutch policy transformations in response to increasing ethnic diversity in Dutch society, especially in cities. This is then tied to the case study of Amsterdam's Bijlmermeer. The case study focuses on the context of the urban renewal of this neighbourhood, and provides a background for the discussion of the negotiation of difference in the production of cosmopolitan and non-cosmopolitan space.

While multiculturalism until the 1990s was central to public policy in the Netherlands, a series of events have recently contributed to the decline of multiculturalism as a policy and its replacement by a more assimilationist stance towards immigrant integration (Entzinger 2003). The Dutch government did not begin to develop a coherent policy to promote immigrant integration until the early 1980s. Before, public policy was based on the assumption that immigrants would only stay temporarily. The report *Ethnic Minorities*, published by the Scientific Council for Government Policy (WRR) in 1979, was very influential in the policy shift. This report stated that immigrants were staying permanently and that public policy was needed to facilitate immigrant incorporation (WRR 1979).

The Council's recommendations were formalized in the *Minderhedennota*, a policy document accepted by the Dutch parliament in 1983. This formed the institutional framework for multicultural policy. Under this policy, immigrants were perceived 'in terms of their group membership and not primarily as individuals, as in most other European countries' (Entzinger 2003: 62). It recognized that the Netherlands had become a multicultural society in which everyone should have equal opportunities. Under this policy, integration was demanded but without immigrants having to give up their cultural identities. The policy, however, did not apply to all ethnic minorities, but only to those groups that the government felt needed assistance, mostly those from the former colonies and non-European guest workers. Other groups, for example the Chinese or immigrants of European origin, were believed to have integrated and therefore were not included under this policy (Entzinger 2003: 63). Hence, the state plays an important role in defining what is perceived as 'Dutch'.

In the 1990s, the government recognized that multiculturalism had not significantly improved the conditions of migrants in the Netherlands. At the same time, ongoing immigration had added to the diversity amongst migrants, thereby rendering the focus on groups in multicultural policy problematic. Therefore, in a new policy document, the *Contourennota* (1994), the individual immigrant became the subject of integration policy, and integration itself was given a more two-sided character. Multiculturalism was not replaced and immigrants could still retain their cultural identity. However, more emphasis was placed on the responsibility of the immigrant to succeed in Dutch society. A Minister of Integration Policy was appointed in 1998 to monitor the process of integration. Another responsibility of the minister was urban policy, and it was in this context that disadvantaged urban neighbourhoods become associated with immigrants and the problem of integration.

While the position of migrants in Dutch society did not improve much over the years, a cosmopolitan ideology held by the Dutch elite prohibited critiquing this situation. It was not until 2003, in a post-9/11 context of increasing unease with immigrants, that immigration was brought to the centre of political debates. In 2003 parliamentary research was demanded to investigate the successes and failures of immigrant integration policies (*Onderzoek Integratiebeleid* 2003). Following the report of the commission, the government formalized a new, more repressive, stance on immigrant integration. This stance is very similar to the position that Mitchell (2004: 645) describes, in which 'those who choose not to assimilate are represented as individuals unwilling to participate in civic life who can, as a result, be excluded from society without incurring damage to the core ideals of a universalist liberal project'.

Integration is no longer voluntary. Instead, the new approach demands assimilation into Dutch society and does not recognize difference as such. A nationalist element – the idea that there is a set of norms and values that are inherently Dutch – has returned in policymaking, and has replaced the more cosmopolitan ideology associated with the former multicultural policy. However, what is most significant is that the problem of integration – defined by mostly white planners and politicians – has become spatialized. The problem of integration is associated with certain urban neighbourhoods, particularly neighbourhoods with large immigrant populations like the Bijlmermeer. The solution, therefore, is urban renewal, geared towards attracting the middle and upper classes to the neighbourhoods, or the dispersal of low-income groups over city regions. In order to attract middle-income residents to disadvantaged neighbourhoods, new marketing strategies have to be developed. Marketing diversity can be seen as one strategy to attract new groups, thereby turning the problem – immigrants – into an asset – diversity.

What does this policy discourse mean for the Bijlmermeer? A central strategy for the revitalization is to diversify the neighbourhood characterized by a low-class migrant population by attracting middle-class residents. The Bijlmermeer is one of Europe's leading examples of a high-rise estate. It is a neighbourhood located in the Southeast district of the city of Amsterdam, on the outskirts of the city. The district has a population of approximately 82,000 residents, of which 46,000 live in the Bijlmermeer (Dienst Onderzoek en Statistiek 2004a, 2004b). Over 70 per cent of the population of the Bijlmermeer is of non-Dutch origin: 30 per cent of the population are Surinamese, 5 per cent Antillean, 10 per cent Ghanaian, and the other 25 per cent have origins elsewhere in the world (see Table 9.1), together making up its

Table 9.1 ***Population of Southeast by ethnicity (%), 2000***

	Surinamese	*Antillean*	*Turkish*	*Moroccan*	*South Europeans*	*Other non-industrial*	*Ghanese*	*Ethnic minorities*		*Industrial countries*	*Dutch*
								2000	*1993*		
Amsterdam	10	2	5	8	2	9	1	36	30	10	55
Southeast	31	6	1	2	2	13	7	62	53	5	33
Bijlmer-Centre	40	9	2	2	2	16	10	81	72	3	16
Bijlmer-East	33	8	1	2	2	16	10	72	63	4	24

Source: Aalbers *et al.* 2003: 87

highly multicultural population of 130 different nationalities. The Bijlmermeer was constructed between 1960 and the mid-1970s, and over the years it has become associated with various social problems, such as crime, poverty, pollution, illegal immigration, social isolation and a negative image. In this respect, the experience of Bijlmermeer is not that different from other large housing estates in Europe that have experienced similar problems of deprivation (Aalbers *et al.* 2003; Helleman and Wassenberg 2004). Rather than being a cosmopolitan space, the Bijlmermeer is constructed externally as a non-cosmopolitan space, a view which fails to recognize the inherently cosmopolitan lifestyles of its multicultural residents, who are negotiating difference on a daily basis through neighbourhood interactions as well as in their connectedness to transnational networks. Its location on the periphery of the city, its poor and ethnic population, and the social problems in the neighbourhood have rendered it a place to avoid rather than a place to visit in order to experience different foods, colours and smells. The migrant resident of this neighbourhood has become identified as the 'Other', and in terms of crime, drug abuse and illegality, rather than in terms of the richness of his/her culture, and its transnational connections.

Cosmopolitan space, on the other hand, is more associated with consumption and is visited when one 'feels like' experiencing difference, rather than from a desire to learn from other cultures. Difference in these spaces is idealized and isolated from the socio-economic conditions of the 'Other'. The Bijlmermeer urban renewal programme aims to re-image the district as cosmopolitan by drawing on its diversity (Stadsdeel Zuidoost 2002). Could cosmopolitanism serve to make this space less threatening, as it did in Manchester's Gay Village where the description of the area as 'cosmopolitan' made it more accessible for a straight audience (Binnie and Skeggs 2004)?

Before exploring this, some insight into the historical development of Bijlmermeer is needed in order to understand its current problems.

How did this spiral of decline occur in a neighbourhood that was only built four decades ago? Bijlmermeer was designed according to the ideas of Le Corbusier and the Congrès Internationaux d'Architecture Moderne movement on modern living. These architects and city planners propagated the functional separation of living, working and recreation (Mentzel 1990; Helleman and Wassenberg 2004). In the Netherlands these ideas were central to city planning in the 1950s and 1960s. Underlying these planning practices was the idea that social life in neighbourhoods could be managed through physical planning. The plans for the Bijlmermeer were very ambitious. It was believed that the Bijlmermeer would become the 'City of Tomorrow' – a modern, innovative place for the middle class to live in. The neighbourhood was built to accommodate 90 per cent of the residents in uniform high-rise blocks with large-scale green, open spaces between them. Roads and railways were elevated so that ground level became the domain of pedestrians and cyclists (Helleman and Wassenberg 2004). Workplaces were located on the other side of the railroad tracks. The high-rises were characterized by large semi-public corridors connected to the parking garages, the idea being that people would meet and interact in these spaces. The idea of collectivity was central to the design, thereby ignoring an increasing individualization in society.

The Bijlmermeer never became the 'City of Tomorrow'. According to Helleman and Wassenberg (2004) three categories of problems are believed to have caused the failure of the high-rise neighbourhood. First, the plans were never fully realized because of financial problems. The apartments were built but the amenities that were promised to the residents – stores, metro-line and spaces for sports and recreation – were never delivered or only arrived years later. Second, the high-rises had to compete with the new single-family homes in the suburbs that became available at the same time. The middle classes preferred the single family homes in the new suburbs to the high-rise apartments, leaving the new flats empty. To avoid vacancies the apartments were rented to starters on the housing market, and when many Surinamese families moved to the Netherlands following the independence of Surinam in 1975 the availability of apartments in the Bijlmermeer directed them to this neighbourhood. Immigrants from Surinam and the Dutch Antilles were soon followed by other immigrants, causing the Bijlmermeer to become the 'gateway' for many newcomers, adding to the multicultural 'flavour' of the district. A final set of issues which contributed to liveability problems related to the management of buildings and public and semi-public spaces. The dark parking garages, corridors, abundant green spaces and tunnels were perceived as unsafe. No one took responsibility for the green public spaces, and residents disposed of their garbage in the quickest manner, by throwing it from their balconies. Today social problems are believed to be the main cause of the design failure (Helleman and Wassenberg 2004). The semi-public spaces were taken over by litter, the homeless and drug dealers.

A further problem is the socio-economic marginalization of the Bijlmermeer's residents. In the 1980s and 1990s the residents of the Bijlmermeer were increasingly isolated from opportunities in the urban economy. The lack of skills, experience and alienation from the political system excluded many residents from Dutch society. High unemployment, debt problems and other social problems led to the further deterioration of the neighbourhood. Until the mid-1990s less than 5 per cent of the total population of Bijlmermeer worked in the adjacent business centre, which provided employment for over 65,000 people at the time, and is amongst the most profitable business districts in the nation. The Bijlmermeer became a more single-class, low-income and unemployed, ethnically diverse, and increasingly non-white enclave (Bair and Hulsbergen 1993), and an entry point for immigrants, both legal and illegal, from all over the world. The Bijlmermeer became a place to avoid, rather than a place to visit to experience and consume the culture of the 'Other'.

After series of renovations and small improvements to the high-rises, in 1992 the state, the city of Amsterdam, the Southeast district council and the housing corporation finally decided on the large-scale urban renewal of the Bijlmermeer. These plans were also strongly supported by the business district which was affected by the negative image of the Bijlmermeer. This operation involves the demolition of over 13,000 homes, approximately 50 per cent of all the high-rises, between 1992 and 2010. These will be replaced with 15,000 homes in low-rise or four-story blocks. The anonymity associated with the green spaces, dark garages and corridors will have disappeared after the renewal operation. Before, 92 per cent of the apartments in the Bijlmermeer were in the social rental sector; after the renewal operation this number should be reduced to 55 per cent (Kwekkeboom 2002), reflecting the ambition of the city to attract middle-class residents to the neighbourhood and to allow people in the Bijlmermeer to make a housing career within the neighbourhood. Besides the physical renewal there are also funds allocated, on a smaller scale, for socio-economic renewal through the European Union and the Dutch Big Cities Policy Programme.

The plans are very ambitious. In 2010 the neighbourhood should have become what it was intended to be when it was designed in the 1960s, the 'City of Tomorrow'. As stated in a promotional brochure produced by the City District marketing the district to new residents and to attract new businesses, 'everybody involved is committed to making the Southeast district truly the "City of Tomorrow" this time around' (promotional brochure, Stadsdeel Zuidoost 2002). The promotional strategy for the neighbourhood centres on the ethnic diversity of the neighbourhood, reflected in the slogan 'The Colourful Perspective of Southeast'. Whereas ethnic diversity of the district was previously equated with social deprivation, the promotional strategy now focuses on the strengths of the existing diversity:

> The Southeast District derives its unique character and dynamics from the exceptional blend of peoples and cultures. This cultural diversity has injected an enormous wealth of insights, knowledge, skills and talents into the district. It is one of the main reasons why people are keen to live and work in the Southeast district.
>
> (Stadsdeel Zuidoost 2002)

Ethnic diversity as the guiding characteristic of the neighbourhood is promoted as a positive aspect, despite the current political climate in the Netherlands that is sceptical of the position of immigrants in Dutch society, and despite the strong national assimilationist stance towards integration rather than multiculturalism. In this sense the revitalization of the district is geared towards cosmopolitan consumption, attracting new middle-class residents to the neighbourhoods who would benefit from the wealth of difference that is present in the neighbourhood. The district is also described in promotional literature as 'Southeast, a vibrant community!' or 'colourful and dynamic Southeast'. And Bijlmermeer is believed to have become the vibrant heart of this multicultural community. The idea of multiculturalism is retained but it is also stressed that the Bijlmermeer is characterized by its social cohesion, based on the number of organizations, churches and arts venues. It is a multicultural community where diversity is rendered ordinary. Multiculturalism is part of the daily life of its residents which has now been taken on as a marketing strategy to attract new middle-class residents, and to broaden its appeal to a wider audience.

Negotiating cosmopolitan urbanism

The goal of urban renewal in the Bijlmermeer was to differentiate its housing in order to attract a wider range of income groups and to allow residents to follow a housing career within the Bijlmermeer. The state (city and district government), the housing corporation and capital – in the form of investments from major corporations with their headquarters located there – were especially interested in rejuvenating the area, especially after it was decided that Southeast was designated to become the second entertainment centre of Amsterdam. Residents, on the other hand, initially resisted the plans for the demolition of 50 per cent of the high-rises. The reason for this opposition was a perception of top-down planning by 'white' technocrats which prioritized physical renewal rather than addressing the social and economic problems prevailing in the neighbourhood. In addition residents blamed those governing the district for 'deporting the blacks' to make the Bijlmermeer more 'white' (Boer 2003: 8). In 1996, this caused a major rift between the 'white' governors and the 'black' residents. The discussion was launched by a group of primarily Surinamese Dutch who had united into the 'Zwart Beraad' (Black Consultation). They pointed out that the white population was in the minority (only 25 per cent) in the Bijlmermeer but the political administration, management and consultants where white (Kwekkeboom 2002). From that moment onwards ethnic organizations were given a more prominent role in the renewal process and top positions within the local council were reserved for ethnic minorities so that the council was more representative of its multicultural population.

Also, from the perspective of the governors and the business centre, Southeast is to become 'a multicultural theme park for city and region' (Kwekkeboom 2002: 84). The business organization (representing the collective interests of the firms in the business centre), for example, exhibits the art work of local artists and invests in local cultural initiatives. The city, using funds for social and economic renewal, supports ethnic art and the organization 'Imagine, Identity and Culture', a project that encourages migrants to visualize their personal stories and present these to a wider audience.

But despite these efforts, the image of the Bijlmermeer as a 'ghetto', a neighbourhood associated with drug abuse, illegal immigrants and social deprivation, still persists in the public imaginary. In order to change this negative image, a group of residents of various ethnic backgrounds have taken the initiative to promote their neighbourhood in terms of the positive elements of multiculturalism. They established the organization 'United Different Voices' – representing the 130 different nationalities present – to promote the neighbourhood from the viewpoint of its residents. United Different Voices organizes fieldtrips in the Bijlmermeer, showing people around the sites where renewal is in progress, introducing them to ethnic organizations such as churches, and having them participate in cultural practices such as cooking Surinamese food or African music and dance. But who then attend these fieldtrips? The visitors are mostly white, elite cosmopolitan consumers, who get to experience the 'exotic' for a day, after which they return to their homes. The participants are mostly college graduates, employees from the adjacent business centre, politicians, and policymakers from Holland as well as from abroad.

For the visitors this experience is one to remember, but mostly one that can be associated with cosmopolitan consumption of the immigrant 'Other'. What stood out for the visitors was the exotic: 'smells and colours', 'foreign food products', 'it is as if you are in a different country', 'only coloured people', and 'the Eastern stores' (United Different Voices 2004). Recently, the local council has incorporated this strategy by establishing a tourist board for the district to attract visitors, and to realize the idea of the Bijlmermeer as a 'multicultural theme park'.

While diversity is stimulated, the number of places where diversity can be experienced, in the form of restaurants and theatres, is still limited. This is not surprising, as many residents of the neighbourhood have limited resources to open up businesses. Rents in the new apartments are high and not suitable for starting entrepreneurs. In many ways the Bijlmermeer is still a non-cosmopolitan space, but a space where the most 'exotic', as Tuan (1996) has argued, can still be found. Despite these changes it is not yet commodified for cultural consumption. However, as Tuan (1996) suggests, there is a danger in opening up such non-cosmopolitan spaces to become 'multicultural theme parks', as this introduces the dangers of commodification and homogenization. Would this be desirable for the Bijlmermeer?

Conclusion

Ethnicity and class are important categories for defining what is considered cosmopolitan or a cosmopolitan space. In this chapter we have seen that cosmopolitanism is not restricted to the elites, who are targeted for experiencing and consuming difference, but is lived on a daily basis by multicultural communities, and even promoted by them in an effort to revitalize the image of the neighbourhood. Stigmatized as a ghetto and as the location of criminality and illegality by outsiders, residents of the Bijlmermeer do live cosmopolitan lifestyles. Cosmopolitanism out of necessity rather than by choice is what distinguishes their cosmopolitanism from that associated with 'white elites'. In their everyday life they are forced to negotiate difference in their interactions with the multicultural community.

This chapter has had two aims. On the one hand it seeks to show that in today's ethnically diverse societies multiple cosmopolitanisms exist and are not restricted to the elite and intellectuals who have the means to consume difference, but are also present in the lifestyle of working-class ethnic 'Others'. Second, the chapter suggests that while on the one hand urban policy and governance are focused on attracting spending power to the city by appropriating difference, at the same time this approach can be a strategy for the ethnic 'Other' to represent their neighbourhood as a part of the city to challenge perceptions of it as an 'external' ethnic ghetto. In this way multiculturalism is negotiated on the ground, through the actions of such neighbourhoods in making their presence felt in society rather than suffering from social and spatial exclusion.

However, despite efforts to market the neighbourhood in terms of its cosmopolitanism, and despite the initiatives of residents to express its cosmopolitanism, the Bijlmermeer seems to remain a space for the non-cosmopolitan 'Other': an 'Other' that is denied cosmopolitanism, despite daily lived experiences with difference and connectedness to transnational networks. The new Bijlmermeer has attracted an ethnic (mainly Surinamese) middle class. In 2015, 81 per cent of the population of Bijlmermeer will consist of immigrants, compared to 43 per cent in the city of Amsterdam. These predictions indicate that the neighbourhood will continue to function as a gateway to the Netherlands for many new immigrants. While these people might start at the bottom of the social ladder, the neighbourhood's inherently cosmopolitan character, and the goals and efforts of residents to communicate this to the outside world, cannot be ignored.

The Bijlmermeer, with its multicultural population, is at the heart of current debates about difference in Dutch society. In many ways this neighbourhood can be seen as the 'City of Tomorrow', because not only is Bijlmermeer the site of the largest urban renewal operation in the country, it is also the place where difference is negotiated and a multiethnic society created. Multiculturalism is the future for cities in the Netherlands, Europe and elsewhere in the world. Therefore, the conflicts and struggles that are ongoing in the Bijlmermeer might become symptomatic for other places. Hence, we might want to reconsider our image of the Bijlmermeer and regard it as a neighbourhood from which we can learn and find solutions for the management of difference in multicultural cities.

References

Aalbers, M.B., Beckhoven, E. van, Kempen, R. van, Musterd, S. and Ostendorf, W. (2003) *Large Housing Estates in the Netherlands. Overview of Developments and Problems in Amsterdam and Utrecht*, RESTATE 2nd report, Utrecht: Faculty of Geosciences.

Bair, T.L. and Hulsbergen, E.D. (1993) 'Designing renewal on Europe's multi-ethnic urban edge. The case of Bijlmermeer, Amsterdam', *Cities*, 6: 283–98.

Beck, U. (2002) 'The cosmopolitan society and its enemies', *Theory, Culture and Society*, 19: 17–44.

Binnie, J. and Skeggs, B. (2004) 'Cosmopolitan knowledge and the production and consumption of sexualized space: Manchester's gay village', *Sociological Review*, 52: 39–61.

Boer, J. (2003) 'Nieuwe Bijlmerm moet zich nu bewijzen', *Nul 20 Tijdschrift voor Amsterdams Woonbeleid*, pp. 8–11.

Caglar, A. (2002) 'Media corporatism and cosmopolitanism', in S. Vertovec and R. Cohen (eds), *Conceiving Cosmopolitanism. Theory, Context, and Practice*, Oxford: Oxford University Press.

Contourennota Integratiebeleid Etnische Minderheden (1994) Tweede Kamer der Staten-Generaal, 1993–94, 23 684, Nos 1–2.

Dienst Onderzoek en Statistiek (2004a) *Kerncijfers Amsterdam 24*, Amsterdam: Stadsdrukkerij Amsterdam.

Dienst Onderzoek en Statistiek (2004b) *Stadsdelen in Cijfers 24*, Amsterdam: Stadsdrukkerij Amsterdam.

District Southeast (2002) *Towards Zuidoost: The Colourful Perspective of Amsterdam Zuidoost*, Amsterdam: Sterprint Grafische Partners.

Entzinger, H. (2003) 'The rise and fall of multiculturalism: the case of the Netherlands', in C. Joppke and E. Morawska (eds), *Toward Assimilation and Citizenship: Immigrants in Liberal Nation-States*, New York: Palgrave Macmillan.

Featherstone, M. (2002) 'Cosmopolis: an introduction', *Theory, Culture and Society*, 19: 1–16.

Hannerz, U. (1990) 'Cosmopolitans and locals in world culture', *Theory, Culture and Society*, 7: 237–51.

Harvey, D. (2000) 'Cosmopolitanism and the banality of geographical evils', *Public Culture*, 12: 529–64.

Helleman, G. and Wassenberg, F. (2004) 'The renewal of what was tomorrow's idealistic city. Amsterdam's Bijlmermeer high-rise', *Cities*, 21: 3–17.

Kwekkeboom, W. (2002) 'Rebuilding the Bijlmermeer 1992–2002. Spatial and social', in D. Bruijne, D. van Hoogstraten, W. Kwekkeboom and A. Luijten (eds), *Amsterdam Southeast. Centre Area Southeast and Urban Renewal in the Bijlmermeer 1992–2010*, Bussum: TOTH Publishers.

Le Galès, P. (2002) *European Cities. Social Conflict and Governance*, Oxford: Oxford University Press.

Mentzel, M. (1990) 'The birth of Bijlmermeer, 1965: the origin and explanation of high-rise decision making', *Netherlands Journal of Housing and Environment Research*, 5: 359–75.

Minderhedennota (1983) Tweede Kamer der Staten-Generaal, 1982–83, 16 102, Nos 20–21.

Mitchell, K. (2004) 'Geographies of identity: multiculturalism unplugged', *Progress in Human Geography*, 28: 641–51.

Onderzoek Integratiebeleid (2003) Tweede Kamer der Staten-Generaal, 2003–4, 28 689, Nos 8–9.

Robbins, B. (1998a) 'Introduction part I: actually existing cosmopolitanism', in P. Cheah and B. Robbins (eds), *Cosmopolitics: Thinking and Feeling beyond the Nation*, Minneapolis, MN: University of Minnesota Press.

Robbins, B. (1998b) 'Comparative cosmopolitanisms', in P. Cheah and B. Robbins (eds), *Cosmopolitics: Thinking and Feeling beyond the Nation,* Minneapolis, MN: University of Minnesota Press.

Rogers, A. (2000) 'Citizenship, multiculturalism and the European city', in G. Bridge and S. Watson (eds), *A Companion to the City,* Oxford: Blackwell.

Sandercock, L. (1998) *Towards Cosmopolis: Planning for Multicultural Cities*, New York: Wiley.

Sandercock, L. (2003) *Cosmopolis II: Mongrel Cities of the 21st Century*, London, New York: Continuum.

Stadsdeel Zuidoost (2002) *Towards Zuidoost: The Colourful Perspective of Amsterdam Zuidoost*, Amsterdam: Sterprint Grafische Partners.

Tuan, Y. (1996) *Cosmos and Hearth: A Cosmopolite's Viewpoint*, Minneapolis, MN: University of Minneapolis Press.

United Different Voices (2004) Available at www.amsterdamzuidoost.net (accessed 4 February 2005).

Vertovec, S. and Cohen, R. (eds) (2002) *Conceiving Cosmopolitanism: Theory, Context, and Practice*, Oxford: Oxford University Press.

WRR (Scientific Council for Government Policy) (1979) *Ethnic Minorities*, The Hague: WRR.

10 Working-class subjects in the cosmopolitan city

Chris Haylett

Introduction

In popular usage 'cosmopolitan' expresses a modern style of urbanity characterized by cultural liveliness and a certain sophistication. Its symbols are chic cafés, arts festivals, international fashion and food, and vibrant streetlife. In academic usage 'cosmopolitan' is transposed to 'cosmopolitanism' where it has political and ethical content, offering an idealized view of society as a place of 'togetherness' where 'otherness' has been banished to less civilized times. This ideal is structured around cultural openness to a mix of differences, primarily those of 'race' and ethnicity, expressed in terms of a connectedness beyond boundaries of national identity, a 'planetary humanism' which includes 'the otherness of the other' (see Gilroy 2000; Beck 2002). Neither popular nor academic versions of 'the cosmopolitan' present us with a view of the urban in which social class is a significant feature; instead class is superseded by culture. The language of cosmopolitanism does not readily conjure images of the black or white working class, or of poor immigrants or refugees. Its central ethic or 'good' is cultural openness and tolerance, not social and economic equality. This chapter argues for the importance of seeing cosmopolitanism in relation to class, an importance which can be explored through *discourses* of urban culture and identity, and *practices* of urban life in the dense flows, mixes, conflicts and convergences of social difference. First, with regard to how urban culture is represented, how we assess the usefulness of cosmopolitanism as an idea largely depends on our positionality or 'way of seeing'. This is suggested by the question 'Who gets to articulate the nature of *our* contemporary urban conditions?' Given the historical and contemporary development of Western cities through class relations, analyses of urban culture that do not consider class are missing a fundamental dimension of urban space and its narration – very often expressing a non-reflexive middle-class positionality. Second, with regard to practices of urban life, the recognition that cosmopolitanisms represent the best way of conceiving of urban cultures, which are too heterogeneous to be lumped together in a single analytic (Pollock *et al.* 2000), requires accounts of cosmopolitanism to be grounded in terms of place and class relations, that is, in historical and geographical conditions.

In this chapter I critically approach urban cosmopolitanism as a discourse rather than an empirical fact, but recognize that its deployment as a discourse has material effects through which meaning and value are attached to certain subject positions and places. So, for example, certain types of people and types of city can be marked as 'cosmopolitan' and the idea of cosmopolitanism can be mobilized in empirical research, in theory and in urban governance (for example, in city marketing) in relation to certain people and places. In other words, ontological realities can be interpreted and rendered in terms of cosmopolitanism. Ulrich Beck, for example, discerns the ontological realities of 'our times' in major shifts of identity, politics, business, work, consciousness and time-space relations, and defines them as globalization and cosmopolitanization, the latter being the internal dynamic of globalization or 'globalization from within the national societies' (Beck 2002: 17). Whilst it is possible to discern an increasing porosity of national boundaries and an increasing density and pace in global flows of culture, money, and politics as described by Beck, there are of course other possibilities of interpretation. My approach to cosmopolitanism is to regard it as a discourse which does not adequately describe working-class conditions of urban life or hold promise as a political project for working-class groups. However, to the extent that it is deployed as a discourse in ways that have a bearing on working-class people in cities, it should be engaged with. In this chapter my intention is to show the descriptive and political limits of cosmopolitanism when it is explored through class and gender-based experiences of neoliberal welfare governance in an apparently 'cosmopolitan' city. Through this empirical case study the importance of social welfare ethics – or civic welfarism – to wider considerations of cosmopolitanism will be argued, and it will be suggested that in cities riven by class-based inequalities, civic welfarism is a preferable discourse of analysis and understanding.

The chapter will be structured into five main parts: first, a brief discussion of the absence of class thinking in many formulations of urban culture; second, the introduction of Houston as a classed 'cosmopolitan city'; third, discussion of the conditions of work and welfare in the United States through which Houston's urban cultures are formed; fourth, an in-depth case study of a culturally oriented welfare reform programme through which relations between class and cosmopolitanism in Houston are analysed; and, fifth, analysis of the political grounds of cosmopolitanism in relation to those of civic welfarism. Throughout the chapter, by using place and politics as the coordinates of my argument, I want to show how centrally class matters to assessments of cosmopolitanism.

Class culture

Class cultures represent a large part of the 'difference' that makes cities what they are – heterogeneous sites of people, meshed together through economic, cultural, political and social processes and practices, marked by diversity and inequality. However, much of the literature surrounding urban culture has preferred to work with concepts of difference focused on 'race', gender, ethnicity, religion, youth and sexuality as framing analytics (Bhabha 1994; McRobbie 1994, 1999; Rutherford 1990; Hall 1992; Miles *et al.* 2000). Where class is named, it is most commonly evoked as an 'old' form of social relation displaced by the 'new', or it is contained in contingent notions of inequality or exclusion related to employment and social opportunities (Giddens 1994). It is rarely considered as *class culture*. This reflects a traditional social scientific understanding of what class is – usually economic structure or position, which has profound consequences for how class can be thought.[1] Certainly the effort of critical thinking that has gone into notions of 'race', gender and sexuality has not been extended to considerations of class culture and identity. Apart from the holes in social understanding that exist because of this, it also serves a particular kind of class-based but class-silent politics in which credentials can be sought for 'new' political projects by middle-class groups positioned at the vanguard of change. In other words political credibility is claimed without having to seriously confront issues about class society and classed selves in relation to classed 'Others'. The liberal discourse of cosmopolitanism, extended as a political project, can be seen as an example of such manoeuvring (see Haylett 2001).

Although academic writers concerned with questions of ethics and justice generally seek distance from uncritical notions of cosmopolitanism, there is also much continuing support for this inclusive ideal (see Held 1995). There are some good reasons for this, most obviously the critique of ethnocentric and territorial notions of identity and place which can lead to colonialisms and parochialisms of various kinds. However, what is put in the place of these 'backward' notions of cultural difference often bears little relation to realities on the ground, through which these notions of identity make sense to people.

This chapter will consider how the urban reality of 'poor work' and welfare disentitlement for large numbers of people creates a need for security that cosmopolitanism does not address as it prioritizes issues of cultural openness and mobility over those of social protection and stability. I will argue that cultural openness is encouraged and enhanced by conditions of social and economic security, and that social welfare should therefore be seen as a conceptual and political condition of a cosmopolitanism based on equal rather than exploitative cultural exchange. Urban culture is the medium through which class struggles for welfare are lived on a daily basis and is full of ontological struggles for respect, dignity and value. These might only be resolved by recognizing how social and cultural conflicts are related to unequal economic and geopolitical power: that is, how disrespect and inequality are usually of a piece.

Houston – what kind of city?

So how do urban conditions of work and welfare in 'advanced' economies relate to cosmopolitanism? Practices of work and welfare constitute the daily grind of urban life for the vast majority of people living in and around contemporary cities. Since the 1970s these cities have been undergoing momentous changes in their physical form, economic base, political governance, social structure and cultural life. Central to these have been changes in employment and welfare entitlement as the Fordist economic order and its social contract is seen to have passed. De-industrialization has brought about the decline of 'older' industrial cities in Western Europe and North America, sometimes to re-emerge as new urban economies, in competition with dynamic growth in new urban regions. Houston, in the south-east of Texas, is one such city-region regarded as riding the tide of late capitalism, through economic growth in its leading service and retail sectors (Scott and Soja 1996). It is also a city marked by racial and ethnic diversity, with large populations of Hispanic (37 per cent), African-American (25 per cent) and Asian (5 per cent) people, in a total city population of 1.8 million (as of the 2000 census). The city government draws on this rich diversity to promote the city's image in cosmopolitan terms through its e-Government centre and public relations material (see www.cityofhouston.gov). Houston has two Chinatowns and a Little Saigon (with the second-largest Vietnamese population in the United States); the largest performing arts district outside of New York, with 90 groups devoted to multicultural and minority arts; a prestigious fashion retail and design centre (the Galleria); annual ethnic festivals (Greek, Egyptian, Japanese); and a high number of ethnically diverse residential areas, made famous both through rap (South Park), and through comparison with Beverly Hills (River Oaks). However, to the extent that Houston has grown as a prosperous city-region, this growth is striated by class differences and inequalities. Houston's diverse urban cultures do not simply rub 'alongside' its prosperity, making this economy a 'city of difference' as well as a city of economic growth. Beyond easy promotional and encyclopaedic representations of 'the cosmopolitan city' (see, for example, nationmaster.com) there are social processes at work, which yoke culture to economy and work them *through* each other.

Culture and economy in Houston are thoroughly intertwined, and because the social processes that intertwine them are of a neoliberal capitalist nature, they produce inequality. This inequality is lived through urban cultures, which are the medium of daily life. None of this culture *can* be untouched by class because class relations are the constitutive dynamic of the capitalist social processes that make urban space 'work' (Harvey 1973). It is therefore significant that class is not always named or explored in relation to urban culture's forms and practices of identity and sociality. Certainly in the United States it has been a historical achievement to displace class as a primary category of social understanding and to replace it with differences of 'race' and ethnicity in particular (see Aronowitz 1974, 1996). In Houston the engineering project of making class invisible works physically through the cityscape in the creation, segregation and concealment of working-class residential areas within toxic industrial zones filled with refineries and chemical plants such as Baytown and La Porte (Houston is the largest city in the United States without zoning laws); and in the creation of bypass zones, such as the Bayou, crossed by freeways that dominate the city form and provide its logic. The project of erasing class visibility also works discursively by occluding class-based understandings of the city. The governance of Houston, like the rest of Texas, is ultra-neoliberal, characterized by a tradition of one-party Republican politics and business dominance. Houston is the economic centre of Texas's petrochemical industries, its port and airports are amongst the busiest in the world, and international labour flows constitute its workforce at all levels (Vojnovic 2002). In this political economic context, we might ask what the naming of Houston as a cosmopolitan city achieves, and for whom? For example, is there an argument for cosmopolitanism to address the class inequalities and social ruptures of Houston? My suggestion is that Houston *can* be analysed from the perspective of cosmopolitanism in terms of its 'culturally rich urban life' and potential for cultural openness and conviviality, and from the perspective of neoliberalism in terms of the competitiveness of its urban cultural assets including its 'cosmopolitan feel', but that neither perspective properly engages with the working-class city of Houston. A much more pointed description, politics and vision of the working-class city of Houston is found in the idea of civic welfarism which engages with the main front of attack against Houston's working-class poor (of all 'races', ethnicities and genders), namely the withdrawal of social value and security from them via welfare restructuring. These people constitute the borderlands of national legitimacy: a no-man's-land between welfare and work, a place where their citizenship – literally their right to the city of Houston – is effectively withdrawn. How does this place of internal exile relate to cosmopolitanism?

Houston is a world city, a massive node for transnational flows of capital, culture and people, which means that in Beck's terms it can be figured as a pre-eminent example of globalization and cosmopolitanization (Beck 2002). It is also pre-eminently divided, and not just in terms of economic fortunes as might be expected. Houston's economic divisions *work through* social and cultural spaces in ordinary daily social practices and in the intimate embodiments of class. These intimate geographies might be explored through the bodily effects of living as working-class people in the most polluted city in the United States with absolutely minimal environmental protection and healthcare entitlement. Or through the bodily effects of doing the daily rounds of city life, through rump public transport, housing and welfare services. Or through the embodied nature of encounters in the local and federal institutions through which poor working-class groups are contemptuously processed, from prisons to food pantries, shelters, missions, community colleges, social service and employment agencies. The nature of these working-class geographies is only partially expressed (if at all) in statistics or theories of diminishing forms of 'collective consumption', or in rare accounts of 'ordinary cosmopolitanisms', more straightforwardly described as examples of anti-racism (see Lamont and Aksartova 2002). This partiality suggests the value of the Foucauldian concept of 'bio-politics': of social power that works in capillary form to colonize bodies, internally and externally, to capture the deeply material, subjectivity-making effects of class relations working through bodies, minds and selves (Foucault 1977; Harvey 1996). By taking the analysis of urban culture to this level, cultural difference can be class-specified: what *kinds* of difference mark differently classed bodies? How do these differences require differential thinking? What does class difference do to ways in which 'difference' is evoked in notions of cosmopolitanism? This level of analysis disallows the possibility of talking about cultural diversity 'alongside' and 'as well as' economic inequality in a liberal 'here we all are' approach to cosmopolitan cities. Some further elaboration is needed here.

The intertwining of culture and economy through class and space is not neutral or innocent. What is readily lost from cosmopolitanism's idealization of coexistence and retreat from the idea of 'incommensurable' differences between people is the conflict inherent in living within class-based social relations. If this conflict is suppressed it is possible for cosmopolitanism to 'appear' in cities but not to 'be' in any socially or politically progressive sense of the word. Unfortunately, much of the liberal discourse of cosmopolitanism and the subject position it instantiates – 'the cosmopolitan' – is happy about and even dependent on things not going much beneath what 'appears' to be the case. In a world figured in terms of multiple and overlapping cultural differences the constitution of 'the cosmopolitan' through class-based subjectivity is not disclosed. By way of example, cosmopolitanism is easily claimed from transitory middle-class contact with 'Others' in their service capacity as newsagents, cleaners and waitresses, or from visual contact with 'Others' who make up the colourful backdrop to city life as it is passed through as an authentic backdrop in gentrifying areas (see May 1996). However, beyond these performances a more personally and socially involved 'openness' is likely to involve more fundamental conflicts and shifts in exchange relations from which substantive benefits for 'Others' might be achieved. Here social welfare is primary as the route by which minimal goods of citizenship are represented and distributed: welfare states institutionalize social relations of consensus and conflict which test the limits of openness and commitment to 'Others'. Conversely, suppressing the discourse and social relational conflicts of class, or pronouncing society 'classless' or class-confused, makes civic welfarism much less likely as an intellectual project, or political and popular ideology. This has been evident in academia and politics since the 1980s when the headlong retreat from welfarism began. Its effects can be seen if we study urban cultures of work and welfare more closely, in the hidden and not so 'hidden injuries of class' which mark the bodies, minds and selves of working-class women, men and children (Sennett and Cobb 1972).

Urban cultures of work and welfare in the United States

The momentous changes shaping North America's urban landscapes over the past 25 years or so have involved withdrawing employment opportunities and welfare entitlements on a grand scale, a process that has been greatly heightened by welfare reforms in the past decade (Peck 2001). Among other social effects, conditions of social reproduction or the daily work of sustaining life have been undermined by the withdrawal of 'social contracts' or 'social wages' for activities outside of paid work such as the raising of children (Piven and Cloward 1997; Mink 1998). The restructuring of America's welfare state therefore involves more than restructuring particular forms of work and community – it involves remaking urban economies and classes at a structural level. It also has effects of a deeply personal and political kind in domestic and intimate spheres of social reproduction, in relationships between partners, parents and children for example, and between so-called 'dependent' and 'independent' groups in society more widely. This consideration is particularly relevant to ideas of cosmopolitanism which are interested in extending openness, concern and support to 'Other' groups in need of respect and/or resources. Where conditions of social reproduction are radically undermined by social welfare systems so that citizenship is effectively denied to large groups of working-class people, then the chances of openness being extended to 'dependants' (of various kinds, economic and otherwise) from other countries are all the more remote. This is less problematic for transnational elites of course, whose money and power buy access and 'independence'.

The idea of civic welfarism is about cultural and economic support for social relationships between differently classed groups based on the recognition of interdependence and intersubjectivity between societies and selves. Here there is no 'outside': splitting from the 'Other' is splitting the self (see Rustin 1991). In contrast to this ideal, 'workfare' – or compulsion into the labour market through the withdrawal of benefits for childrearing – closes possibilities of citizenship for citizens and immigrants alike. The institutional practices of workfare deploy a discourse of self, 'Other' and society based on personal responsibility, individualism and the closing down of collective solidarities. This discourse works through signifiers of class, 'race' and gender to mark bodies as more or less legitimate depending on their immediate economic utility. In this culture of citizenship, 'togetherness' is found in market relations of consumption and production, and for the poor, social value only comes through labour market participation. The social significance of these imperatives of welfare subjectivity are manifold. Here we can consider them for what they say about relations between class and culture and welfarism and cosmopolitanism in the governance of Houston. In particular we should consider the ways in which cosmopolitanism can be a useful discourse for neoliberalism.

Family Pathfinders

Houston's workfare programmes all operate within the federal structure of welfare reform instituted in the 1996 Personal Responsibility and Work Opportunity Reconciliation Act. This act allows individual states to implement their own levels, limits and conditions of welfare entitlement for mothers with children, the only group eligible for any subsistence-level help. The ethos of Texas's welfare reform is 'work first' based on lifetime limits for benefit receipt (1–3 years), personal responsibility (no benefit increases for the birth of additional children), and the immediate placement of participants in employment (as opposed to training or education). In Texas, teenage parents and immigrants are not eligible for temporary benefits at any point, and the majority of welfare recipients are African-American women with children.

Houston prides itself on being a greatly philanthropic city (Kelley 2004). It is also a city where lines between government, business and religion are blurred, and this unholy trinity is celebrated. Those occupying the higher echelons of these fields are Republicans and evangelical Christians. They belong to fundamentalist right-wing Protestant churches whose goal is to 'remoralize' America (see Olasky 2000). As part of this mission there are fundraisers and benefactors of programmes orientated to reduce welfare rolls and supply low-wage labour markets. A prime example is the flagship Family Pathfinders programme, which provides transitional support for Houston families being moved from 'welfare-to-work'. It has three main parts: a job search programme made up of 'soft skills' or 'life skills' run by the Texas Department of Human Services; a job training programme organized by the Texas Workforce Commission which represents low-wage state employers in retail, secretarial, customer service and computing sectors; and 'coordinated community involvement' from volunteer teams drawn from local churches, businesses and civic organizations – primarily charities such as Lions Clubs. The programme was set up in 1996 and is now processing 170 families at any one time, taking referrals from the Department of Human Services and churches across Houston. In 1997 Family Pathfinders was recognized by a White House welfare-to-work symposium as a model worthy of replication throughout the United States. Variants of it are already widely established throughout the United States (Bartkowski and Regis 2003).

The conjunction of 'soft skills' and job training elements with 'coordinated community involvement' is claimed to be Family Pathfinders' point of innovation. This marks a specifically cultural engagement with families: bringing families into a relationship with volunteers and a wider social space beyond households and neighbourhoods. This is in keeping with the wider discourse of welfare reform, pitched to reform so-called cultures of 'welfare dependency' which are represented as an endogenous condition of working-class poverty and its ghetto-like existence in housing projects and other urban enclaves. Family Pathfinders is conceived as a way of helping 'transitional welfare families' into new ways of personal, family and working life. It is structured to bring together 'different' sections of the community *to have a relationship*. Although not stated in class terms, the basis of this difference is class and the basis of the 'exchange' is class-cultural. Families are meant to find personal and practical support and benefit from contact with the values, behaviours and cultures of role models. Volunteers are meant to develop compassion and understanding for poor families, which can be carried back into community lives. There is no financial support given. The idea is to develop the cultural capacities of families to find and keep work, by giving them cultural knowledge about the world of work, developing their self-esteem, and giving them role models to aspire towards and emulate.

Family Pathfinders operates amongst a wider range of volunteer, exchange and church-based community schemes, which are central to how US culture imagines itself 'bridging' cultural and economic divides (Warren 2001). These schemes are very often the only point of contact between 'different' social groups beyond service relationships. A central aspect of how they work is by actively constructing relationships across social and cultural differences. However, in a historical and geographical context these relationships are not simply about 'cultural exchange'. To the extent that programme participants are able to benefit from contact with middle-class cultures, these benefits are highly prescribed (by the needs of employers, for example). Moreover 'benefit' rests on the assumption that middle-class forms of cultural capital can be imported into, make sense within, and be beneficial to, working-class cultural worlds. This is to imagine that working-class people can be more like middle-class people given the right kinds of instruction and example. Here there is a fundamental misunderstanding of culture as a 'good' which can be mechanistically imported into people's lives, rather than a way of life constituted through social, economic and political conditions. In urban and institutional conditions of class inequality, such as prevail in the operational context of Family Pathfinders, what we have instead of cultural mixing and 'exchange' is cultural exploitation, through which unequally classed subjectivities are produced.

The understanding of culture as process brings attention to the discursive production of both the cosmopolitan and the working-class subject, in relation to each other and the wider cultural space. Among the discourses of self and society deployed in Family Pathfinders are those of the American nation, neoliberal capitalism, and liberal identitarianism. The shared coordinates of these discourses are self-advancement and the work ethic; Christian values and beliefs; and ethnic, racial and gender difference. These discourses are manifest in a wide variety of ways in the institutional practices of Family Pathfinders, and in the performances of the women being processed. The common currency of particular ways of talking is perhaps most indicative. Predominant among the volunteers, administrators and the women themselves are various kinds of religious and therapy 'speak' taken from self-help counselling, conservative evangelicalism, consumer culture, and cheap confessional television shows that make up daily media diets. This 'speak' takes its most explicit form in group classes and ceremonial events where it is actively encouraged.

Here exhortations of self abound: 'You can make it!' 'Go for it!' 'Don't let him hold you back!' 'You owe it to yourself!' 'Move on!' 'It's your life!' 'God will help you!' 'Do what *you* want to do!' 'You deserve it!' 'Raise yourself up!' 'No-one can do it for you!' 'Go girl!' 'Respect yourself!' 'It's your decision!' These performances were observed as part of a research project funded by the Royal Geographical Society entitled 'Welfare and Culture in Texas and Louisiana: Exploring Links Between Policy and Identity', in August–September 2001. This language reflects a wider culture of personal 'life-management' through which mantras of personal responsibility, aspiration, motivation and self-actualization become norms of social and self-understanding (see Rose 1992).

The project of Family Pathfinders to create 'self-determining' women is one that strongly involves positive identitarian discourses of 'race', gender, sexuality and religion as energizing impulses of self-empowerment and actualization. Performances and articulations of black sisterhood, evangelical 'self-raising', reproductive control, liberation through work, positive self-image and sexual expressiveness convey readily available forms of popular feminism and multiculturalism which are an integral part of the programme's rationale to produce governable welfare subjects. All that is missing from these rainbow identifications is any expression of class identity. Positive expressions of racial, gender, sexual and religious identity are emptied of class identity – as the identity which gives these other markers 'negative' social content. Class is therefore made silent, painful and shameful, while race, gender, sexuality and religion are made positively performative, non-political and conflict-free. The lack of usefulness of these positive identitarian discourses to the women in personal and political respects can be explored through this negation of conflict.

Family Pathfinders is not a poverty-alleviation programme. At its best it might be considered an 'anti-othering' programme, but one that is based on a culturally unequal exchange, as the programme manager expresses it: 'in terms of lasting results we do more for the volunteer teams than we do for the clients' (Programme Manager, Family Pathfinders, personal interview, 27 August 2001). This could be seen as positive in terms of an appreciation of how 'Other' lives are led. Inter-community *relations*, social *contact*, cultural *associations*, anti-othering *exchanges* and positive *expressions* of racial, ethnic and gender identities can all be claimed by Family Pathfinders. But the class-cultural context of the contact suggests that it merely reinforces a power relation in which accomplished middle-class selves are further dignified through the shame, inadequacies and gratitude of working-class 'Others'. From another view, it works as cultural exploitation of working-class women, logically connecting to their economic exploitation. This logic is rolled out through the city, where female migrant workers maintain professional women in paid employment, while conditions of stratified familism mean that poor women are disallowed the status or possibility of raising their own children (Ehrenreich and Hochschild 2002). What therefore does it mean to say that Houston is a 'cosmopolitan city' when its welfare governance is based on profound class inequalities and class-cultural exploitation?

Cosmopolitanism or civic welfarism?

Cosmopolitanism in Houston, as in many other 'advanced' urban economies, contains elements of multiculturalism and identity politics which are mobilized in programmes of urban governance such as Family Pathfinders. Whilst this chapter does not argue that there are no valuable forms of encounter taking place between different cultural groups on the programme or in the city, the question remains of what is gained by the working-class poor from this programme, or from a political framing of their lives in terms of cosmopolitanism? All the women on Family Pathfinders are working-class, most of them are African-American, some of them are white, and all of them are racialized. Through widespread discourses of 'underclass' and 'white trash' those 'on welfare' are positioned as a race or breed apart from American citizens. This is a crucial point to consider because it reveals how the working-class poor – black and white, male and female – are subject to inseparable processes of classification and racialization, marked as abject beings who have fallen from 'the nation' (see Haylett 2001). A mistake of cosmopolitanism is to imagine that racialization is the primary process of cultural division. The women on Family Pathfinders are not impoverished and exploited because of their 'race' but because their dependence on welfare means they can be racialized as *apart from the nation*, unable to fulfil American's destiny to succeed and 'work'. So when 'race' cross-cuts working-classness, poverty often follows, but it does not often follow when 'race' cross-cuts middle-classness. The problems of these women will not therefore be solved by cultural openness and mixing which is always marked by class and makes unequal cultural resources and 'selves' available to different class groups. For example, the middle-class 'cosmopolitan self' is produced and positioned to develop cultural capital, whereas the working-class self is more readily produced and known through discourses which are about a lack of value, such as those of dependency, backwardness, worklessness and illegitimacy. Moreover, the 'cosmopolitan self' has working-class 'Others' as its necessary backdrop: objects available for the cultivation of a superior self.

In Houston, class cultures are inextricably interwoven with virulent class inequalities produced by the distributions of neoliberal systems of work and welfare. The liberal discourse of cosmopolitanism is not cognizant of, interested in, or equipped to deal with these class cultures and inequalities. In fact liberal cosmopolitanism can be actively useful in reinforcing the social injustices produced by those systems, as shown by the identitarian discourses of the possessive individual self reiterated on Family Pathfinders. Moreover, where cosmopolitanism is underlain by a liberal or culturalist politics it is singularly destructive of the very possibility of social solidarity. This leads us to consider the ways in which social solidarities might be forged, and the possibilities of resolution already with us. For example, cultural forms of class struggle are expressed in popular socialist forms of social welfare, in the representations and distributions of child benefits, minimum income guarantees, pensions, national health and social services, for example. These have traditionally expressed a will to social security and solidarity between genders, generations and classes, and across cultural differences. Although in practice welfare states have, at different times and places, discriminated against women and ethnic groups (Williams 1995), they have also been the best chance women and ethnic groups have had to reduce their economic inequality, and have been used as a strategic resource to this effect (Mink 1998). Through socialist and feminist impulses for redistribution and care (Fraser 1989), welfare has been able to extend notions of citizenship through intersubjective notions of the self, or notions of the self in which connections to proximate others are part of the self, the 'I' that is 'We'. This is the basis of civic welfarism (also see O'Neill 2001).

Historically in the United States, and currently in Europe, Australia and New Zealand, welfare states have proved to be the best way of recognizing and moving towards the reduction of social conflicts and cultural closures. More than just an institution of state, welfarism works as a political ideal and practice in a way that can create social solidarity through both representational and distributional means. Its capacity to do this is derived from its 'will-to-class-struggle' which recognizes and works to reduce grounds of social conflict which are inherent to capitalist societies, in precisely the way cosmopolitanism does not in its effort to build consensus from the 'contingency of identity'. Liberal cosmopolitanism's promotion of concepts of contingency and consensus over welfarism's focus on equality and conflict reveals both political interests and weaknesses of liberalism, wishfulness and elitism. It also translates to a lack of geographical specificity, where grounded realities of history and power take vernacular form.

To the extent that the United States – 'a nation of immigrants' – might be seen as 'cosmopolitanism's source' (Robbins 2002: 32), it is an instructive case study in what can go wrong when discourses of class and welfare are displaced by those of cultural identity and diversity. In common with every other US city, welfare reform in Houston is based on culturally oriented programmes like Family Pathfinders which are perfectly able to enmesh positive identitarian discourses of identity based on 'race', gender and sexuality with the racialized segregation of the working class into conditions of forced labour. The transnational role of the United States, leading the way in destroying notions of welfare that are not about compulsion into the labour market (Peck 2001), suggests that we need to engage normative arguments against this state of affairs at the global level where policy directions are increasingly determined. Notwithstanding the necessity of this global perspective, some argue that the role of *national* welfare states is still crucial in creating solidarity with people perceived to be different from oneself (Hollinger 2000). This argument forwards the idea that human beings feel the need to belong and act at smaller spatial scales than 'the planet', recalling Gilroy's description of cosmopolitanism as 'planetary humanism'. Others insist that only a *transnational* project of welfare, foregrounding issues of economic inequality, would be a 'left cosmopolitanism' worth its name (Robbins 2002).

Either way, these considerations of cosmopolitanism's scale position welfare as central to concerns about social solidarity. In this chapter I have argued that the logic of this positioning favours civic welfarism over cosmopolitanism as a discourse and practice of urban living and justice. The case study of Houston discussed here suggests that civic welfarism is best able to address both ontological and ethical crises of social security and solidarity, up against those who can say with impunity of working-class women struggling to raise children in straitened circumstances, 'We want them to work more, earlier and harder' (Advocacy Director, Family Pathfinders, personal interview, 23 August 2001).

Note

1 The work of Pierre Bourdieu (1984) is an important exception here, bringing class-cultural formulations to the fore of social theory, though he has had much less 'use' than theorists working with sexuality, 'race' and gender, such as Michel Foucault, Stuart Hall and Judith Butler. From the late 1990s sociologists started to address matters of class culture more generally, and there now exists a small number of strong exceptions to the rule (see Skeggs 1997, 2004; Reay 1997; Munt 1999).

References

Aronowitz, S. (1974) *False Promises: The Shaping of American Working Class Consciousness*, New York: McGraw Hill.

Aronowitz, S. (1996) *The Death and Rebirth of American Radicalism*, New York: Routledge.

Bartkowski, J. P. and Regis, H. A. (2003) *Charitable Choices: Religion, Race, and Poverty in the Post-Welfare Era*, New York: New York University Press.

Beck, U. (2002) 'The cosmopolitan society and its enemies', *Theory, Culture and Society*, 19: 17–44.

Bhabha, H. K. (1994) *The Location of Culture*, London: Routledge.

Bourdieu, P. (1984) *Distinction: A Social Critique of the Judgment of Taste*, London: Routledge.

Ehrenreich, B. and Hochschild, A. (2002) *Global Women: Nannies, Maids and Sex Workers in the New Economy*, New York: Metropolitan Books.

Foucault, M. (1977) *Discipline and Punish: The Birth of the Prison*, London: Penguin.

Fraser, N. (1989) 'Struggle over needs: outline of a socialist-feminist critical theory of late capitalist political culture', in N. Fraser, *Unruly Practices: Power, Discourse and Gender in Contemporary Social Theory*, Minneapolis, MN: University of Minnesota Press.

Giddens, A. (1994) *Beyond Left and Right: The Future of Radical Politics*, Cambridge: Polity Press.

Gilroy, P. (2000) *Against Race: Imagining Political Culture Beyond the Color Line*, New York: Harvard University Press.

Hall, S. (1992) 'The question of cultural identity', in S. Hall, D. Held and T. McGrew (eds), *Modernity and its Futures*, Cambridge: Polity Press.

Harvey, D. (1973) *Social Justice and the City*, London: Edward Arnold.

Harvey, D. (1996) 'The body as an accumulation strategy', paper presented at the annual conference of the Association of American Geographers, North Carolina, April 1996.

Haylett, C. (2001) 'Illegitimate subjects?: abject whites, neoliberal modernization, and middle-class multiculturalism', *Environment and Planning D: Society and Space*, 19: 351–70.

Held, D. (1995) *Democracy and the Global Order: From the Modern State to Cosmopolitan Governance*, Oxford: Polity.

Hollinger, D. A. (2000) *Postethnic America: Beyond Multiculturalism*, New York: Basic Books.

Kelley, M. (2004) *The Foundations of Texan Philanthropy*, College Station, TX: Texas A&M University Press Consortium.

Lamont, M. and Aksartova, S. A. (2002) 'Ordinary cosmopolitanisms: strategies for bridging racial boundaries among working class men', *Theory, Culture and Society*, 19: 1–25.

McRobbie, A. (1994) *Postmodernism and Popular Culture*, London: Routledge.

McRobbie, A. (1999) *In the Culture Society: Art, Fashion and Popular Music*, London: Routledge.

May, J. (1996) 'Globalization and the politics of place: place and identity in an inner London neighbourhood', *Transactions of the Institute of British Geographers*, 21: 194–215.

Miles, M., Hall, T. and Borden, I. (2000) *The City Cultures Reader*, London: Routledge.

Mink, G. (1998) *Welfare's End*, Ithaca, NY: Cornell University Press.

Munt, S. (ed.) (1999) *Cultural Studies and the Working Class*, Basingstoke: Cassell.

Olasky, M. (2000) Foreword by George W. Bush, *Compassionate Conservatism: What it Is, What it Does, and How it Can Transform America*, New York: Free Press.

O'Neill, J. (2001) 'Oh, my others, there is no other!: civic recognition and Hegelian other-wiseness', *Theory, Culture and Society*, 18: 77–90.

Peck, J. (2001) *Workfare States*, New York: Guilford Press.

Piven, F. and Cloward, R. (1997) *The Breaking of the American Social Compact*, New York: The New Press.

Pollock, S., Bhabha, H. K., Breckenridge, C. A. and Chakrabarty, D. (2000) 'Cosmopolitanisms', *Public Culture*, 12: 577–89.

Reay, D. (1997) 'Feminist theory, habitus and social class: disrupting notions of classlessness', *Women's Studies International Forum*, 20: 225–33.

Robbins, B. (2002) 'What's left of cosmopolitanism?', *Radical Philosophy: A Journal of Socialist and Feminist Philosophy*, 116: 30–7.

Rose, N. (1992) 'Governing the enterprising self', in P. Heelas and P. Morris (eds), *The Values of Enterprise Culture: The Moral Debate*, London: Routledge.

Rustin, M. (1991) *The Good Society*, London: Verso Books.

Rutherford, J. (ed.) (1990) *Identity: Community, Culture, Difference*, London: Lawrence and Wishart.

Scott, A. and Soja, E. (eds) (1996) *The City: Los Angeles and Urban Theory at the End of the Twentieth Century*, Berkeley, CA: University of California Press.

Sennett, R. and Cobb, J. (1972) *The Hidden Injuries of Class*, New York: Vintage Books.

Skeggs, B. (1997) *Formations of Class and Gender*, London: Sage.

Skeggs, B. (2004) *Class, Self, Culture*, London: Routledge.

Vojnovic, I. (2002) 'Governance in Houston: growth coalitions and environmental and social pressures', paper presented at the annual conference of the Association of American Geographers, Los Angeles, March 2002.

Warren, M. (2001) *Dry Bones Rattling: Community Building to Revitalize American Democracy*, Princeton, NJ: Princeton University Press.

Williams, F. (1995) 'Race/ethnicity, gender and laws in welfare states: a framework for comparative analysis', *Social Politics: International Studies on Gender, State and Society*, 2: 127–59.

11 Planning Birmingham as a cosmopolitan city: recovering the depths of its diversity?

Wun Fung Chan

Introduction

In the last 20 years, the city of Birmingham in the UK has been rebranded into a forward-looking regional centre that is embedded within Europe and the global economy. The first phase of Birmingham's reinvention involved the construction of flagship developments, such as the National Exhibition Centre, the International Convention Centre and Symphony Hall. These prestige developments, although instrumental in enhancing Birmingham's international profile, have been heavily criticised for being spatially and socially out of reach for many local residents (Lister 1991; Fretter 1993; Smyth 1994; Duffy 1995; Loftman and Nevin 1996). However, with the second phase of Birmingham's reimagining the local authority has sought to involve a much wider constituency through building a new ethos for the city (Bhattacharyya 2000). This has included a self-conscious effort to involve and reflect the city's cultural diversity in the regeneration of what has officially been described as a *cosmopolitan urban environment*:

> We will maintain and develop Birmingham's status as a world class and cosmopolitan urban environment, business location and visitor destination; and continue to create and sustain the condition for economic diversification and growth, reflecting the city's multi-cultural communities.
>
> (Birmingham City Council 2002: 131)

To consider how such a cosmopolitan urban environment could possibly be built, a conference called 'Highbury 3' was held in

February 2001. Made up of planners, politicians, voluntary sector workers, local business people, academics and representatives from Birmingham's ethnic minority communities, the conference featured a number of discussions on healthy living, communication links, the provision of education, neighbourhood management, the new economy, crime, the urban environment and cultural development. Each of these discussions was, in the words of Councillor Bore, 'to meet the challenge of diversity' and sought to establish ways in which the city could peacefully manage its cultural differences. To quote one of the key sound bites scattered amongst the conference proceedings:

> One of the key concepts explored at Highbury 3 was the fact that the city's great strength and defining characteristic is its *depth of diversity* . . . Managing this diversity in a positive way is *the challenge facing the city* – to exploit the richness it offers and avoid the potential conflict and tensions that could arise from lack of connectivity between people and communities, and between communities and the 'city'.
>
> (Birmingham City Council 2001: 12, emphasis in original)

Amongst the many participants was Charles Landry of Comedia, whose organisation was commissioned by Birmingham City Council to produce a research report to assist further dialogue. Entitled *Planning for the Cosmopolitan City*, this report was composed by the highly respected urban commentators Jude Bloomfield and Franco Bianchini (2002). In it, they refine and elaborate upon many of the sentiments contained in *Highbury 3* such as: 'making the most of assets presented by the city's many different communities', 'forging a vital bond between the city's entrepreneurial tradition and its new ethnic diversity', '[living] interculturally' and 'welcoming . . . all groups from within and visiting the city'.

In this short chapter, I outline Birmingham's plan for a cosmopolitan city and, moreover, point towards some of the issues and agendas that are in need of attention if Birmingham is to embark on such a project. The two main questions I ask of it are: how deep is the city going to dig to recover the 'depth[s] of [its] diversity'?; and is the introduction of cosmopolitanism in Birmingham's cultural life something new or is it an iteration of certain values? To seek to answer them I detail the recent proposals that seek to re-plan Birmingham as a cosmopolitan city and excavate one of its key features, the idea of ethnic minorities as an asset, through urban regeneration policy and the planning of Birmingham's Chinese Quarter. By drawing upon Jacques Derrida's (2001, 2002) deconstruction of hospitality and extending David

Parker's (2000: 77) 'discourse of cultural contribution', I demonstrate that Birmingham's proposals, although seemingly quite innovative, contain a number of contradictions that mark some limits to how a cosmopolitan urban environment is being cultivated.

In a collection such as this one it would seem appropriate for each of its contributors to define their take on the terms in question: cosmopolitan urbanism. However, this chapter neither sets out the parameters of cosmopolitanism, nor does it theorise the slippery relationship between cosmopolitanism and urbanism. Others have already embarked upon such tasks in this volume and elsewhere. Rather, my intention is to critically think through how cosmopolitanism has become delimited and solicited by policymakers and their advisers in an urban context: Birmingham. The means through which I make sense of this city's complex narrative on cosmopolitanism is by engaging with Derrida's consideration of hospitality, albeit somewhat implicitly. For the sake of exposition, Derrida's notion of hospitality can be understood as located within what he calls 'a hidden contradiction between hospitality and invitation' (Derrida 2002: 362). This refers to the way hospitality runs counter to and in competition with an invitation that may be offered under duress, or to fulfil a debt, or out of legal or moral obligation. As Derrida goes on to explain, such an invitation signals a type of pact, a failure to escape the familiar, and as such it marks a limit perverting the possibility of a 'welcome'. Thus, hospitality can only be understood as a welcome that extends beyond names, invitation or expectations. In this chapter, I follow a similar current to suggest that, whilst there is evidence in Birmingham's planning discourse of a welcome orientated towards incorporating the city's ethnic minorities in the city's rebuilding, it is possible to locate the elision of a number of other subject positions once the terms of the 'welcome' are exposed.

Reading *Planning for a Cosmopolitan City*

Britain's second largest city, Birmingham, has long been associated with a modernist ethos. This is perhaps no better reflected than in the legacy of the post-war planning projects that left the city's skyline with a number of large-scale, monolithic and concrete buildings. Built upon a foundation of entrepreneurial endeavour, manufacturing and technological innovation, the city sat at the centre of Britain's industrial revolution both geographically and metaphorically. Indeed, famed for the production of cycles, aircraft, munitions and motorcars, Birmingham has proudly been called 'the workshop of the world'. Nonetheless, with the harsh recession of the 1970s, the city experienced widespread labour shedding and reductions in reinvestment in the engineering trades leaving many jobless and rendering areas of the city as unproductive.[1]

To reverse the city's economic decline, Birmingham embarked upon an ambitious attempt to reorientate the local economy towards the service sector. The National Exhibition Centre was opened in 1976. Investment was sought in areas such as office spaces, upmarket retailing and in the leisure industries through a strategy of remarketing Birmingham's image. Flagship developments were also built – such as the International Conference Centre, the Hyatt Hotel and Brindley Place – to provide an infrastructure that could support corporate and prestige events. In addition, the city centre was reorganised and redecorated around six 'quarters' in the late 1980s, one of which was based around a 'Chinese theme' (Chan 2004).

Although the construction of Birmingham's Chinese Quarter was to be downplayed with the redevelopment of the neighbouring Bull Ring, the tentative flirtations with issues of cultural difference in Birmingham's planning were placed more firmly on the agenda with *Highbury 3*. In a sense, this was a long time coming. Birmingham's population is characterised by its post-war Commonwealth immigration and, despite having been marginalised in the city's politics, ethnic minorities and migrants have long been a spectral presence overlooking the rebuilding of the city (Chan, in press). As such, Birmingham's attempts to meet 'the challenge of diversity' marks something of a turning point in its planning agenda and deserves close attention. In this section, I highlight some of the main points of *Planning for a Cosmopolitan City*.

Perhaps the key reference point for planning the cosmopolitan city is Leonie Sandercock's *Towards Cosmopolis* (1998, 2003), and certainly the orientations taken by Birmingham's policymakers and advisers are no different in this respect. For instance, the research report by Bloomfield and Bianchini (2002) begins by summarising the main points of Sandercock's thesis in a positive manner. It affirms her appeal for the recognition of 'multiple forms of knowledge of marginalized peoples' and outlines Sandercock's idealisation of a 'city of memory', 'city of spirit' and 'city of desire'. It also recalls the critique of 'Enlightenment epistemology', which dismisses the assumptions of a rational and/or universal planning gaze. Still, despite acknowledging Sandercock's model of 'insurgent planning', Bloomfield and Bianchini seek to draw some distance between themselves and her work. Notably, they critique *Towards Cosmopolis* for failing to conceptualise 'a common public domain with culturally diverse citizens' and for privileging 'multiple publics' (Bloomfield and Bianchini 2002: 5), the outcome of this failure being that Sandercock purportedly understands multiculturalism as a variety of self-contained cultural worlds without intercultural communication or interaction. That a more attentive reading of *Towards Cosmopolis* would note that Sandercock actually and repeatedly refers to 'new kinds of multi- or cross-cultural literacies' (Sandercock 1998: 206) is not my concern here. But what is of interest is the way this critique is mobilised to forward an alternative cosmopolis with its own recommendations and lacunae.

Pivotal to Birmingham's rethinking of cosmopolis are five proposed models that could potentially deal with the city's cultural diversity. These are 'corporate multiculturalism', a 'civic cultural integration model', a 'US melting pot approach', a 'transcultural model' and, finally, 'interculturalism'. On the last of these models Bloomfield and Bianchini (2002: 6) write:

> [The interculturalism] approach goes beyond opportunities and respect for existing cultural differences, to the pluralist transformation of public space, civic culture and institutions. So it does not recognise cultural boundaries as fixed but as in state of flux and remaking. An interculturalist approach aims to facilitate dialogue, exchange and reciprocal understanding between people of different cultural backgrounds. Cities need to develop policies which prioritise funding for projects where different cultures intersect, 'contaminate' each other and hybridise . . . In other words, city governments should promote cross-fertilisation across all cultural boundaries, between 'majority' and 'minorities', 'dominant' and 'sub' cultures, localities, classes, faiths, disciplines and genres, as the source of cultural, social, political and economic innovation.

Interculturalism is a key feature in *Planning for a Cosmopolitan City*. For Bloomfield and Bianchini it is the preferred 'model'. It obtains such a privileged status because intercultural policy, according to the report, may serve to remedy social and economic inequalities 'without reinforcing ethnic divides, by reversing discrimination and exclusion trends' (Bloomfield and Bianchini 2002: 7). The report also goes on to claim that intercultural spaces – such as music venues and community centres – can help counteract ethnic segregation through facilitating cross-cultural collaboration. Nevertheless, by taking interculturalism as a policy to be implemented there is a tendency in the report to presume that the hybrid nature(s) of cities are in need of augmentation. For instance, on the Karnivale der Kulturen in Berlin, which is said to involve 'almost all the minority cultural organizations in the city', the report states, 'the structure of the event is not fully intercultural' (Bloomfield and Bianchini 2002: 6).

By forwarding interculturalism as a model to breach the ethnic divides of communities, the report underplays the fact that cultures have historically become interwoven and inflected to bring about social and cultural transformation (see Hall 1992, 2000). This is an important oversight, because by failing to acknowledge that ethnic cultures are *always already*, for better or worse, negotiating the boundaries presumed between them, the inclination is to treat cross-cultural contacts as something new and unproblematic. In contrast, if cultural contacts are seen to be already in place and a historical feature of city life, then the significance for urban policymakers might be to encourage them to refocus governmental practices away from merely building intercultural relationships per se and instead point them towards a political and ethical questioning of the agendas of such relationships in policy contexts. In particular, it might suggest that policymakers would do well to more carefully map the subject positions that are to be cross-fertilised and the globe-girdling arrangements in which they are situated. One place where they might begin this critical process is through a reinspection of *Planning for the Cosmopolitan City* and a discussion of the other, seemingly tangential, agendas and relationships advanced within it. Interculturalism is not sufficient in and of itself.

A main thread running through *Planning for the Cosmopolitan City* is undoubtedly the concern with issues of separatism and segregation. As previously mentioned, these issues are arranged as the counterpart to the proposals of interculturalism and are deemed to be problematic in as far as that they project an illusion of self-contained, homogeneous ethnic communities. Yet whilst interculturalism is taken as the key remedy to these problems of representation, the report also slips into the discussion a need to instil the citizens of Birmingham with particular rights and duties. As it states: '[ethnic] segregation can only be overcome within a public sphere which confers equal rights and obligations [which] . . . reflects the pluralistic character of its citizens and their cultural make-up' (Bloomfield and Bianchini 2002: 5). Though the report fails to go on to specify what rights and obligations the citizens of Birmingham might actually adhere to, it does feature a number of clues on their potential make-up, some of which are captured in a brief quotation in the introduction:

> How Birmingham responds to demographic and cultural changes will determine whether it becomes a civitas augescens. This is a term used . . . to refer to a dynamic, adaptive city, in which welcoming 'the Others' is a source of strength and imagination. Birmingham has begun to see cultural diversity as an asset and opportunity, rather than a problem or threat, and this is a vital shift in mindset.
>
> (Bloomfield and Bianchini 2002: 3)

This brings us two significant notions concerning the report's intercultural agenda: cultural diversity as an asset and welcoming 'the Others'. On the notion of asset the report indicates a number of ways forward. It writes: '[the] skills and networks acquired informally by ethnic minority young people need to be linked in Birmingham to more formal training, to turn them to economic advantage in self-employment or micro-businesses' (Bloomfield and Bianchini 2002: 8). It critiques Birmingham's official response to the MacPherson Inquiry, *Challenges for the Future* (Birmingham Stephen Lawrence Inquiry 2001), for failing to 'engage sufficiently with entrepreneurial ideas, resources and networks in the city, and with the obstacles to capitalising on them fully' (Bloomfield and Bianchini 2002: 8). It also states that the 'Chamber of Commerce and inward investment agencies need to become much more attuned to the opportunities for transnational networking offered by diasporic business communities – particularly links with India and China' (Bloomfield and Bianchini 2002: 8). The report is quite clear on this front.

In contrast, the notion of 'welcoming the Others' is less well spelled out in *Planning for a Cosmopolitan City*. Admittedly, there is a small reference to civitas augescens; a concept used by the Roman jurisconsult Pomponio to refer to a city that does not withdraw into itself in face of demographic pressures. Adapted by the academic and politician Massimo Cacciari (cited in Amendola 1998: 83), this term has been reinterpreted as a city that recognises the need to 'assimilate peregrinos, hostes et victos' in order to avoid its own decline. But overall the report fails to elaborate upon the city's welcome beyond the introduction. For instance, there is little or no mention of what 'the Others' might obtain, want or need; with the civitas augescens the integration of 'the Others' is seen as a test of the city's strength and opportunity to advance itself.[2] Perhaps more significantly, the report does not consider how an ethic of hospitality might be implicit to the city's welcome, let alone discuss the relationship between the aforementioned asset and otherness. It is to these blind spots that I now turn.

Cultural diversity as an asset

Though Bloomfield and Bianchini go on to look at a series of examples of best practice, what is absent from their report is a critical excavation of their own proposals. In this section I offer such a critique by taking a look at the idea of ethnic minority as an asset. There are a number of points to be made on this. But one way the material might be cut through is by drawing attention to some astute observations made by David Parker (2000) in his paper, 'The Chinese takeaway and the diasporic habitus'. Primarily his paper is concerned with the takeaway as a diasporic space. Following Bourdieu, it examines the way in which everyday intercultural contacts shape the endowments, embodiments and expressions of young Chinese people. Nevertheless, in doing so, Parker also notes that Chinese culture is widely represented through what he calls 'a discourse of cultural contribution' (Parker 2000: 77) whereby this ethnic group becomes represented as a contributor to the dominance of British life. As he keenly observes, the manifestations of this discourse stretch across a variety of fields – including ethnic entrepreneurship, the Queen's welcoming remarks to the former Chinese President Jiang Zemin, the classical racial theory of de Gobineau, as well as academic representations of the overseas Chinese – but beyond the ethnic particularity of Parker's paper it is possible to suggest that this discourse resonates through a number of other registers which position (some) ethnic minorities along the lines of an underexploited resource. In other words, it is possible to note the dispersal of this 'discourse of cultural contribution' in a variety of other narratives. For instance, in former UK Prime Minister Margaret Thatcher's infamous 'swamping' speech she reluctantly accepts the presence of a small number of immigrants as long as they garner the status of an asset to the majority community. As she declares:

> Some people have felt swamped by immigrants. They've seen the whole character of their neighbourhood change . . . Of course people can feel that they are being swamped. Small minorities can be absorbed – *they can be assets to the majority community* – but once a minority in a neighbourhood gets very large, people do feel swamped.
>
> (*Observer*, 25 February 1979, cited in Solomos 1993: 97, my emphasis)

More recently, in a defence of asylum seekers, former Foreign Secretary Robin Cook has mined a similar vein, claiming 'ethnic diversity of Britain is not a burden' but 'is an immense asset that contributes to [Britain's] cultural and economic vitality' (*Daily Mirror* 2001). He goes on to suggest that making the most of these assets is 'a condition of economic success in the modern world'. So although the political terrain and context have altered between these two statements, what remains consistent is a tradition where the presence of immigrants is defended with the registers of contribution and asset as alluded to by Parker's (2000) incisive paper. Exactly what assets they might or could be remains open to conjecture. Still, it is precisely their celebration – what they might add and what they might address – that signals a distinct moment in the recognition of ethnic minorities. The dissemination of this discourse is further evident in the political construction of a highly particularised ethnic subject in British urban policy and local authority planning in Birmingham.

It is well noted that throughout the 1960s to the early 1970s there were a number of changes to UK urban policy, which trace the emergence of ethnic minorities in governmental discourse. Some of the most significant of these shifts occurred with the introduction of the Urban Programme and under Section 11 of the 1966 Local Government Act, which were initially orientated towards social and educational provision, albeit with varying degrees of acknowledgement of race and ethnicity. However, a period between the mid-1970s to the mid-1980s marked a more exaggerated, if not more decisive, turning point in terms of the level of governmental recognition afforded to ethnic differences. For instance, the 1977 White Paper Policy for the Inner Cities acknowledged the presence of ethnic minorities and, following the unrest in 1981, ministerial guidelines were issued by the Department of the Environment (1981) to ensure that 'the special needs of minority groups are recognised and catered for'. Notably, running alongside this appearance of ethnic and racial issues in governmental discourse was another macro-political shift that culminated in an increased valuation of entrepreneurial forms of governance (Harvey 1989). Simplifying somewhat, this shift became underwritten by the hegemonic themes of national duty, self-interest, competitive individualism and anti-statism, which promoted the rolling back of the managerial state, contractualisation, the reconstruction of service provision as an entrepreneurial practice and public–private partnerships. This did little for equal opportunities or attempts to tackle socio-economic disadvantage amongst minority groups (Blair

1988: 49). Yet I would also suggest that the dispersal of the above regulative trajectories, along with the increased recognition of cultural differences, began to share an affinity, if not valorised, a policy formulation of immigrants, or at least some elected features of them, as a potential node of regeneration. Peter Hall's 'free port solution', which sought to 'recreate the Hong Kong of the 1950s and 1960s inside inner Liverpool or inner Glasgow' (Hall 1982: 417), marks one example of this. Another example might be drawn from the way the Urban Programme was increasingly refocused to enhance the employability and business skills of ethnic minorities (Smith 1989; Munt 1994). Returning to the West Midlands, it is also possible to note that the local authority took a number of additional steps to augment these trends in urban policy, albeit on a smaller scale. For instance, in the 1980s attempts were made by the Economic Development Committee to build and support a culture of entrepreneurialism amongst the minority communities and to explore the different avenues that ethnic groups offered in broadening the regional space economy. These measures included the commissioning of various reports such as *The West Midlands Food Industries*, which saw the 'ethnic food market' as a potential means to access overseas markets (Wiggins and Lang 1985),[3] and the *Directory of Ethnic Minority Businesses* (Birmingham Enterprise Centre 1986), which according to Councillor Bore would 'assist inter-trading' and 'benefit the local economy'. In addition, the Business Advice and Training Scheme was established to 'assist ethnic entrepreneurs' in the early 1980s. This particular scheme was the forerunner of Birmingham's Employment Resource Centres whose remit was to 'increase the take-up of business start-up and support services by groups under-represented in business, e.g. people from ethnic minority communities, women and people with disabilities'. Managed by the Birmingham Economic Development Partnership (a joint initiative of the Training and Enterprise Council, Chamber of Commerce and City Council) those funded include the Birmingham Chinese Society.

One of the justifications that underwrites the funding of these organisations is their significance for social inequality. It is well noted that in Birmingham seven of the ten wards with the highest levels of unemployment house 50 per cent of the city's ethnic minorities (Birmingham Stephen Lawrence Inquiry 2001). Yet, unfortunately, social deprivation is not something new to Birmingham or its ethnic minorities. Still, what has begun to appear is a series of programmes in which 'ethnic entrepreneurship' is supported, enlisted and rationalised as a node of regional regeneration and, furthermore, seen as a means to revisualise the city's connection to different international circuits of capital. As the 1997–2000 Economic Strategy for Birmingham suggests, one of the intentions of Birmingham's Economic Development Partnership (1997: 6) would be to create opportunities in the 'multi-cultural industries' so that it was possible to '[build] on the unique advantage that our multi-cultural City gives in international trading arenas'. At one level, this has become focused upon valorising and cultivating a localised ethnic minority individual who would purportedly be self-reliant and have entrepreneurial motivations, yet at another the presence of minorities has also been usefully deployed as a means of reforging Birmingham's relationship with its international hinterlands. The handover of Hong Kong to China in 1997 and its relationship to Birmingham's Chinese Quarter provides one example of the sort of connections that the city has been pursuing.

After the Joint Declaration between Britain and China in 1984 there were growing expectations of both emigration and capital flight from Hong Kong (Lin 1998). In 1994, per capita gross domestic product was higher in Hong Kong than in Britain and Australia (Smart and Smart 1996: 37), and a number of countries, particularly those around the Pacific Rim, reformed their immigration policies to cash in (see Mitchell 1993). In Britain, the governmental response, although not so clear-cut, was one that promoted a conditional form of settlement that was clouded by self-interest. The British Nationality (Hong Kong) Act of 1990, for example, made available a limited number of passports to heads of households and their families according to a points scheme that favoured Hong Kong's corporate, professional, public service and military elite. Whilst much of the governmental discussion surrounding this Act was with rebuilding confidence in the Hong Kong financial market, an upshot was the production of a 'Homo economicus' where citizenship was granted to a select number of individuals for their capacity to participate in circuits of transnational capital (Ong 1998). On a regional scale the construction of this individual became further manifest through the actions of the West Midlands Development Agency who sought to attract the dispersal of Hong Kong's capital-bearing migrants by flagging the very presence of an existing British Chinese population:

> Centred on the flourishing ethnic food industry, the UK's Asian enterprises have moved within a generation from the backstreets of inner city Birmingham to the mainstream of international trade . . . Today the West Midlands Development Agency tries to woo inward investors by

> highlighting the number of Chinese-speaking professionals in the region. Councils are keen to support ethnic businessmen who can provide new jobs in inner city areas. In particular, the region hopes its links with the Far East will attract Hong Kong businessmen [sic] before next year, when the colony reverts to Chinese control.
>
> (*Financial Times* 1996)

In Birmingham the local government response speculated that around 5000 post-1984 Hong Kong migrants would arrive in the region and, moreover, suggested that they would be a source of investment for the city's rebuilding. For these reasons a delegation of officers from the City Council was temporarily located in Hong Kong to solicit affluent migrants and a series of conferences were held between senior city councillors and the Birmingham Chinese Society to identify their potential requirements. In one of them, a spokesperson for the Chinese community echoed the views of Margaret Thatcher on immigration and its assets. More precisely, he stated:

> There will be a lot of confused people arriving here. It could put an enormous strain on existing services. I would like Birmingham to have a policy to encourage people to come here. *They are not poor refugees. They are nearly all professional, well-educated people, many of them with capital to put into starting businesses. They could be an asset to the city.* We are developing a Chinatown in Birmingham as a tourist attraction and it will be a lot more successful if there is a good-sized Chinese community here.
>
> (Steve Yau cited in the *Birmingham Post* 1991, my emphasis)

The marketing of diversity has altered the standing(s) of ethnic minorities in Birmingham. Those with either capital to invest or symbolic value for the tourist industry have become its place entrepreneurs (Lin 1995). Yet whilst these new players in the city's regeneration mark an enculturing of urban policy, ultimately they signal an ambivalent relationship with the differences located both within and outside the city limits. Indeed, as they set in motion ways to defend the presence of the migrant through the idea of an asset, these agents simultaneously reveal a political closure whose syntax cannot help but trace the repression of an emergent heterogeneity. In the above case, the apparent antithesis to the professional, well-educated, wealthy migrant becomes the exclusion of the constitutive figure of the poor refugee. There are undoubtedly other identities amongst them, but one issue that this perhaps leaves poorly addressed is how the ethnic minority as an asset might be reconciled with the city's welcome to the Other.

Conclusion

In the mid-1960s, the *Birmingham Post* published an article entitled 'Cosmopolitania'. Referring to the area surrounding Gooch Street, which was later redefined by planners as part of Birmingham's Chinese Quarter, it depicts a harmonious coexistence between the city's ethnic minorities constructed upon intercultural trade. Indeed, those living around the area are said to exist in a place where 'East meets West', and accordingly the illustrations include a picture of a drapery shop that sells 'material for saris alongside Western cotton shirts' and another of an Indian supermarket whose manager 'finds that his popadoms and curries are as popular with his English customers as the cornflakes and baked beans are with his fellow Punjabis'. The *Birmingham Post* (1967) then goes on to suggest that the area's peacefulness is predicated upon the integration of immigrants into an Anglo tradition:

> [After] 15 years of immigration in Birmingham, the old seemingly inflexible traditions of India, Pakistan and Jamaica are becoming more and more anglicised. The Pakistanis wear trilbies and suede shoes and their children speak like Brummies and talk of Beatles music and mod fashions.

Accompanying this tale of cosmopolitanism is also a brief mention of integration's partner in policy: immigration control. The paper states:

> This year Birmingham's coloured community is feeling the effects of the Commonwealth Immigration Bill. Only 8500 immigrants a year, including 1000 Maltese, will now be allowed into Britain. Many relatives of those already here are in the queue and some face a long wait.

Although dated around 40 years apart, there are a number of parallels with the contents of the article 'Cosmopolitania' and the recent developments in Birmingham. Perhaps most obviously, both cases paint a picture of relative peacefulness, created upon a landscape of intercultural dialogue, entrepreneurial endeavour and shared values. Also apparent and posed as antithetical to the above is a concern with insularity and social tension. Still, given the explicit connotations about cultural difference, migration and settlement, it is surprising to find that there is a reluctance to discuss an ethic of hospitality in both cosmopolitanisms. It would seem all too possible to have cosmopolitanism without hospitality.

For many who have contributed towards the literature on cosmopolitanism there is a concession of an internal ambiguity within its articulation. As Pollock *et al.* (2000: 577) succinctly put it, cosmopolitanism 'escapes positive and definite specification'. Nevertheless, by openly and proudly rebuilding their cities as cosmopolitan entities, city planners and their partners have found ways to cut through the discursive haze. Recent developments in Birmingham for one show how the floating signifier of cosmopolitanism has settled upon a tradition of entrepreneurialism. Redefined as assets, its ethnic minorities have not only followed a tradition that has long underpinned Birmingham's economic development, but marked by their ethnic origin they have found themselves recast as bridges to other worlds. Reflecting the discursive construction of this multicultural subject, the city has also constructed an infrastructure of business forums and resource centres that have been supplemented with the rise of newly defined ethnic quarters. Yet despite these newly found physical structures of ethnic recognition in the city's morphology, there is an inherent tension between the celebration of its minorities as contributors to city life and another of the city's wishes. The notion of welcoming the Other is not something easily reconciled with the above trends.

Within the discourse on cosmopolitanism, there is a line of thought that suggests that 'the cosmopolitan can become a broker, an entrepreneur who makes a profit' (Hannerz 1990: 248). Nonetheless, if this is to become a benchmark in defining the citizens of a cosmopolitan city then it runs against the grain of a welcome to the Other. Whilst notions of asset may act as a useful means to positively re-spin ethnic minority identities from their former associations as a problem, this term is not analogous with soliciting an otherness that an unconditional welcome necessitates. In the above cases, migrants are situated in a broader discourse that equates a right to the city with the exchange of capital contributions and are played out by differentiating the figure of the 'Homo economicus' (Ong 1998) against the figure of an unruly, burdensome migrant. Thus, if Birmingham is to show its welcome, the city might consider showing its hospitality without asking for reciprocation or interrogating a migrant's identity. As such, there is a need to reopen the agenda of Birmingham's cosmopolitan planning to make a home for those on the city's margins.

Notes

1 Between 1971 and 1987, it was estimated that 29 per cent of total employment had been lost in Birmingham (Beazley *et al.* 1997).

2 Cacciari (cited in Amendola 1998: 84) states: 'When a city begins to fear its guest, its weakness, not its strength emerges. In many European countries the opportunity to develop depends on this presence of "the Others".'

3 Later, these reports would be supplemented by the work of the Birmingham Food Forum. This was established in 1994. It initially included five different ethnic minority communities and was supported by Birmingham TEC and Birmingham City Council's Economic Development Department and the Environment Department to promote the food industry in Birmingham. The main aim of the Forum is to develop Birmingham as a multicultural food sector for tourists and local consumers and to provide training in the catering industry.

References

Amendola, G. (1998) *Culture and Neighbourhoods*, vol. 4: *Perspectives and Keywords*, Strasbourg: Council of Europe Publishing.

Beazley, M., Loftman, P. and Nevin, B. (1997) 'Downtown redevelopment and community resistance: an international perspective', in N. Jewson and S. MacGregor (eds), *Transforming Cities: Contested Governance and New Spatial Divisions*, London: Routledge.

Bhattacharyya, G. (2000) 'Metropolis of the Midlands', in M. Balshaw and L. Kennedy (eds), *Urban Space and Representation*, London: Pluto Press.

Birmingham City Council (2001) *Highbury 3: Dynamic, Diverse, Different*, Birmingham.

Birmingham City Council (2002) *Best Performance Plan 2002/2003*, Birmingham.

Birmingham Economic Development Partnership (1997) *Birmingham: The Vision*, Birmingham: Switch Business Communication.

Birmingham Enterprise Centre (1986) *Directory of Ethnic Minority Businesses*, Wolverhampton: Gibbons Barford.

Birmingham Post (1967) 'Cosmopolitania', 17 November.

Birmingham Post (1991) 'City warned of refugee chaos threat', 19 June.

Birmingham Stephen Lawrence Inquiry Commission (2001) *Challenges for the Future: Race Equality in Birmingham*, Birmingham.

Blair, T. L. (1988) 'Building an urban future: race and planning in London', *Cities*, 5: 41–56.

Bloomfield, J. and Bianchini, F. (2002) *Planning for the Cosmopolitan City: A Research Report for Birmingham City Council*, Leicester: Comedia, International Cultural Planning and Policy Unit.

Chan, W. F. (2004) 'Finding Chinatown: ethnocentrism and urban planning', in D. Bell and M. Jayne (eds), *City of Quarters: Urban Villages in the Contemporary City*, Aldershot: Ashgate.

Chan, W. F. (2005) 'Planning at the limit: immigration and post-war Birmingham', *Journal of Historical Geography*, 31(3).

Daily Mirror (2001) 'Rule Tikkannia', 20 April.

Derrida, J. (2001) *On Cosmopolitanism and Forgiveness*, trans. M. Dooley and M. Hughes, London: Routledge.

Derrida, J. (2002) 'Hospitality', in *Acts of Religion*, trans. G. Anidjar, London: Routledge.

Duffy, H. (1995) *Competitive Cities – Succeeding in the Global Economy*, London: Spon.

Financial Times (1996) '£5m pagoda is testimony to Asian success', 29 May.

Fretter, A. D. (1993) 'Place marketing: a local authority perspective', in G. Kearns and C. Philo (eds), *Selling Places: The City as Cultural Capital, Past and Present*, Oxford: Pergamon Press.

Hall, P. (1982) 'Enterprise zones: a justification', *International Journal of Urban and Regional Research*, 6: 416–21.

Hall, S. (1992) 'What is this "black" in black popular culture?', in G. Dent (ed.), *Black Popular Culture*, Seattle: Bay Press.

Hall, S. (2000) 'Conclusion: the multi-cultural question', in B. Hesse (ed.), *Un/Settled Multiculturalisms: Diasporas, Entanglements, 'Transruptions'*, London: Zed Books.

Hannerz, U. (1990) 'Cosmopolitans and locals in world culture', in M. Featherstone (ed.), *Global Culture*, London: Sage.

Harvey, D. (1989) 'From managerialism to entrepreneurialism: the transformation in urban governance in late capitalism', *Geografiska Annaler*, 71b: 3–17.

HMSO (1977) *Policy for the Inner Cities*, London: HMSO.

Lin, J. (1995) 'Ethnic places, postmodernism, and urban change in Houston', *Sociological Quarterly*, 36: 629–47.

Lin, J. (1998) *Reconstructing Chinatown: Ethnic Enclave, Global Change*, Minneapolis, MN: University of Minnesota Press.

Lister, D. (1991) 'The transformation of a city: Birmingham', in M. Fisher and U. Owen (eds), *Whose Cities?* London: Penguin.

Loftman, P. and Nevin, B. (1996) 'Going for growth: prestige projects in three British cities', *Urban Studies*, 33: 991–1019.

Mitchell, K. (1993) 'Multiculturalism, or the united colors of capitalism?' *Antipode*, 25: 263–94.

Munt, I. (1994) 'Race, urban policy and urban problems: a critique on current UK practice', in H. Thomas and V. Krishnarayan (eds), *Race Equality and Planning: Policies and Procedures*, Aldershot: Avebury.

Ong, A. (1998) 'Flexible citizenship among Chinese cosmopolitans', in P. Cheah and B. Robbins (eds), *Cosmopolitics: Thinking and Feeling Beyond the Nation*, Minneapolis, MN: University of Minnesota Press.

Parker, D. (2000) 'The Chinese takeaway and the diasporic habitus: space, time and power geometries', in B. Hesse (ed.), *Un/Settled Multiculturalisms: Diasporas, Entanglements, 'Transruptions'*, London: Zed Books.

Pollock, S., Bhabha, H. K., Breckenridge, C. A. and Charabarty, D. (2000) 'Cosmopolitanisms', *Public Culture*, 12: 577–89.

Sandercock, L. (1998) *Towards Cosmopolis – Planning for Multicultural Cities*, Chichester: Wiley.

Sandercock, L. (2003) *Cosmopolis II: Mongrel Cities of the 21st Century*, London: Continuum.

Smart, A. and Smart, J. (1996) 'Monster homes: Hong Kong immigration to Canada, urban conflicts, and contested representations of space', in J. Caulfield and L. Peake (eds), *City Lives and City Forms: Critical Research and Canadian Urbanism*, Toronto: University of Toronto Press.

Smith, S. J. (1989) *The Politics of Race and Residence: Citizenship, Segregation and White Supremacy*, Cambridge: Polity Press.

Smyth, H. (1994) *Marketing the City: The Role of Flagship Developments in Urban Regeneration*, London: Spon.

Solomos, J. (1993) *Race and Racism in Britain*, London: Macmillan.

Wiggins, P. and Lang, T. (1985) 'The West Midlands Food Industries 1985: a report to the West Midlands County Council Economic Development Committee', London: London Food Commission.

Cosmopolitan knowledge and the production and consumption of sexualised space: Manchester's Gay Village

Jon Binnie and Beverley Skeggs

Introduction

Our approach to this chapter stems from a concern that the mutually constitutive politics of class and sexuality remain undertheorised in contemporary understandings of cosmopolitanism. In many disciplinary spaces class has been edged off the agenda by other differences, namely race and sexuality, as if they can be studied as independent and not mutually engaged categories of constitution and disruption. It is as if the body only carries one sign of difference in its corporeal baggage and that this sign of difference can be separated from the material history of its genealogy and reproduction. The unique contribution of this chapter is to systematically bring together discussions of the politics of class and sexuality within a study of cosmopolitanism grounded in a specific locality – Manchester's Gay Village. We start by giving a brief introduction to theories of cosmopolitanism. What are the main ways we can characterise cosmopolitanism and what have been the main lessons and conclusions from the recent explosion of work in this area? Identifying the key issues in theorising the politics of sexuality, class and cosmopolitanism we proceed to introduce the case study – the production and consumption of cosmopolitan space in Manchester's Gay Village. We explore the different physical and discursive constructions of cosmopolitanism and how they are lived in this particular marked space. Providing a discussion of the context to the study – the development of Manchester's Gay Village – we examine the extent to which the space is branded as cosmopolitan. We then consider the way discourses of cosmopolitanism are deployed by users of the space to enable navigation of that space and the negotiation of difference. We argue that the term cosmopolitan is useful in helping to understand the unease and discomfort with being an appropriate or 'proper' user of the space which requires a fixity of identity, a possession of the right personae to pass through and occupy the space. This argument is developed with particular emphasis on the way heterosexual, 'straight' women consume the space. In asserting their claim on public space not of their own making, we ask whether this demonstrates a rejection of traditional hetero-masculinity and a permeable, transitional space for moving through sexualities, examining the extent to which straight women could be seen to use the space designed specifically for gay male consumption to reformulate the traditional gender relations which they inhabit. We argue that behind and within the articulation and desire for the fluidity of identity associated with the use of the term cosmopolitan, the rigidities of class and lesbian and gay identity are reproduced. In particular, class entitlement plays a major role in articulating and enabling who can be included and excluded from this space.

Sexuality, class and cosmopolitanism

'To some extent, homosexuality represents the last frontier of diversity in our society, and thus a place that welcomes the gay community welcomes all kinds of people' (Florida 2002: 256). This quote by Richard Florida draws us to the key focal point of this chapter, namely how lesbian and gay cultures become configured within debates on cosmopolitanism and urban politics. According to Florida, cities that embrace sexual difference also embrace ethnic and other cultural difference. This contentious and widely popularised claim suggests that gay culture occupies a pivotal role within the production and consumption of urban spaces as cosmopolitan. In this section we discuss competing definitions of cosmopolitanism. How can it be characterised and conceptualised? What are the virtues that are associated with cosmopolitanism, and what vices are the enemies of cosmopolitanism meant to possess? How are the politics of sexuality and class implicated within the politics of cosmopolitanism?

It is not our aim here to provide an exhaustive overview of the rapidly evolving body of literature on cosmopolitanism, but rather to identify some key points that may be useful for thinking through the connections between the class and sexual politics of cosmopolitanism. Clearly how cosmopolitanism is conceived has significant consequences for our ability to theorise these connections. As Bruce Robbins (1998: 12) notes: 'like diaspora, cosmopolitanism offers something other than a gallery of virtuous, eligible identities. It points to a domain of contested politics.' It is evident that the politics of cosmopolitanism are highly contested and any account of them must be partial. It is therefore our intention here to focus on these questions from the perspective of class and sexual politics.

The historical generation of the term cosmopolitanism has substantially diverse trajectories. Most generally it is an oppositional term evoked against all that is fixed, parochial and especially national. It can be broken down in different ways: Delanty (1999), for instance, suggests constitutional patriotism, transnational communities and democratic cosmopolitanism. We are interested in its discursive possibilities to hold together and disrupt different categorisations, how it is used as a disposition of distinction and as spatialisation. The majority of debates about cosmopolitanism are made through four different discourses: anti-national, a type of citizenship, a form of consumption and a form of subjectivity. We place the emphasis on the latter two discourses and how they intervene in the polarised positions represented by Žižek (1997) and Brennan (1997) in opposition to Cheah (1998) and Beck (2000). Žižek and Brennan argue that cosmopolitanism is just the latest cynical way of incorporating differences that are both disruptive and useful to capitalism in its expansion of new markets. In contrast, Beck and Cheah take the more optimistic, liberal stance, arguing that cosmopolitanism opens up new ways of being with others. However, Beck (2000: 100) does note that 'cosmopolitan society means cosmopolitan society and its enemies'; there are always cosmopolitan losers and winners. Central to these different positions is the understanding of culture as something that can be owned by a person in the making of themselves, a way of accruing value to themselves as a cosmopolitan subject. This understanding of culture reflects a version of Pierre Bourdieu's (1986) definition of cultural capital as an embodied disposition (which will be explained in more detail in the following section).

Work on cosmopolitanism tends to emphasise that cosmopolitans are at the privileged, empowered end of what Doreen Massey (1993) has famously termed the power-geometry of time-space compression. Cosmopolitans are seen as being elite professionals in high-status occupations, members of what Leslie Sklair (2001) calls the transnational capitalist class, or what Robbins (2001) terms 'the liberal managerial class'. However, Pnina Werbner (1999: 18) argues that 'even working class labour migrants may become cosmopolitans, if willing to engage with the Other'. And Michael Peter Smith (2001: 108) has taken issue with what he claims is Doreen Massey's construction of the less privileged, claiming that she represents them as 'disconnected victims of global processes, entirely lacking in the dynamic connections to transnational flows that she assigns to place'. Does cosmopolitanism operate to incorporate, de-politicise and make powerless or does it make links between strangers and generate universalism in ways that have not been previously imagined? These stakes in the ability to 'be' cosmopolitan (or inhabit the discursive positioning of the cosmopolitan) are developed throughout the chapter.

Cosmopolitanism is most commonly conceived or represented as a particular attitude towards difference. To be a cosmopolitan one has to have access to a particular form of knowledge, to be able to appropriate and know the Other and generate authority from this knowing. In most definitions cosmopolitanism is not just about movement through culture with knowledge but is an embodied subjectivity that relies on access to the requisite cultural capital to generate the requisite dispositions. For instance, Ulf Hannerz (1996) maintains that: 'cosmopolitanism can be a matter of competence, and competence of both a generalised and a more specialised kind' (1996: 103). This immediately raises questions of education, knowledge, skill and cultural capital. How does one access and acquire these skills? To be cosmopolitan is thus to be educated or sophisticated. It is hard to imagine being thought of as being cosmopolitan without also being sophisticated, and vice versa. To be sophisticated is to be able to present oneself as having a detailed knowledge of world cities, and their culinary delights.[1] In a study of post-war consumption spaces of London's Soho, Frank Mort (1998) notes how the area was represented for the sophisticated: 'Guide books and literary souvenirs played up its exotic and ethnically diverse character, insisting that the quarter could only *truly be understood* by the cognoscenti, or initiated tourists' (Mort 1998: 894, emphasis added). To be sophisticated demands that one has access to the right and appropriate cultural knowledge and dispositions. Yet to speak or name oneself as sophisticated runs the risk of betraying one's own lack of sophistication. In this context Litvak (1997) speaks of '[a]n interest that somehow always betrays itself as a desire; a guilty longing for, or an equally guilty anxiety about, the cultural status that its legitimate possessors or occupants, it would seem, have no need to advertise' (Litvak 1997: 3). Thus to proclaim oneself as cosmopolitan or desiring to become cosmopolitan immediately betrays a certain lack: a certain lack of confidence in embodying cosmopolitanism, an insecure disposition and attitude towards difference. What then are the virtues associated with a cosmopolitan disposition? More saliently for our discussion of exclusions from cosmopolitanism, who are its enemies and its subjects, and what are the vices of cosmopolitanism?

In articulating the virtues associated with cosmopolitanism it is clear that some differences count more than others. Marketing and consumption have a significant role to play in inscribing differences with value (see Tom Frank (1997, 2001) for a brilliant analysis of these value and legitimation processes by which differences are marked). Louisa Schein (1999: 361) demonstrates the imagined cosmopolitanism of experiencing exotic difference through advertising, in which difference is serially produced. Nikolas Rose (2000: 106) also shows how, within city marketing and promotional campaigns, difference 'has been transformed into calculated, rationalised and repetitive programmes'. Frank Mort (1998: 898–9) notes: 'A hybrid version of Franco-Italianate cafe-bar society – serving the ubiquitous cappuccino and espresso – is now available in almost every metropolitan quarter in the world with claims to fashionability.' Banal consumption makes a safe *imagined cosmopolitan* experience of difference available to all; but for the cognoscenti, for those who 'truly understand', which differences matter as Bourdieu and Freud would ask?

In an introduction to *Cosmopolitics*, Bruce Robbins (1998) notes that homosexuals were part of an older cast of cosmopolitan characters. Lesbians and gay men performed a specific role within nationalist discourses, occupying at best an ambivalent position towards the nation-state and signifying a threat through their presence (e.g. state initiatives such as Section 28 in the UK (Smith 1994), and the 'don't ask, don't tell' military policy in the United States) (Stychin 1995; Bell and Binnie 2000, 2002). In Cold War discourses the homosexual male was often represented as a virus – a threat to the boundaries of the safe nation. In contemporary discourses the concept of the 'global gay', it has been argued, can often reproduce banal global consumers, contributing disproportionately to transnational flows of capital (Evans 1993; Altman 1996, 1997, 2001; see Binnie 2004 for an alternative view).

In contemporary framings of cosmopolitanism sexual dissidence has been rendered visible through specific discourses. On the one hand, the rise of the 'global gay' has been celebrated as a form of queer politics – 'We're here, we're queer' – and events are organised around this global association, such as corporate and institutionalised Mardi Gras in Sydney (Markwell 2002) or Europride in Manchester. Moreover, the global gay is incited to consume (places and bodies) through a range of travel guides and adverts in international pink papers and magazines (*Spartacus* is an international bestseller and most tourist guides list lesbian and gay spaces). The internet has enabled further global communication (although usually restricted to English speakers) (Tsang 1996; Wakeford 2002). One danger of the global gay discourse is that it reinforces sometimes homophobic discourses of Western gay men as hypermobile, affluent and privileged consumers. Another danger is that it reinforces ethnocentricity and attaches a convenient label to what are often quite contradictory and complex processes. Peter Jackson (1999) berates those who deploy the term 'global gay' to describe the internationalisation of Western gay identity. He argues that it is a term plagued by ethnocentrism, which reduces indigenous cultures and discourses to specific commercial spaces.

Litvak (1997: 4) argues that gay people – especially gay men – have traditionally functioned as 'objects of such distinguished epistemological and rhetorical aggressions as urbanity and knowingness'. Yet Clarke (1993) notes how lesbians are rarely addressed as consumers, rarely targeted as a specific group and rarely identified as cosmopolitan (except with the short-lived 'lesbian chic'). This, she argues, is because they are not identifiable, accessible, measurable and profitable. Examining the sexual and class politics of cosmopolitanism together is particularly significant, as these are routinely set in opposition to one another – one set of politics are commonly rendered invisible within the other. The urgency of this project is evidenced by the brouhaha associated with Richard Florida's (2002: 258) work:

> The Gay Index was positively associated with the Creative Class – but it was negatively associated with the Working Class. There is also a strong relationship between the concentration of gays in a metropolitan area and other measures of diversity, notably the percent of foreign-born residents.

Similarly, the academic spaces created by queer theory, and the increased visibility and viability of lesbian and gay studies, have also not led to any sustained discussion of intersections of class and sexuality. Whether this is an undertheorisation, an active forgetting or, as Lacan (1977) would argue, 'a will to ignorance' of the work of earlier, less fashionable writers who have in fact studied class alongside sexuality is a moot point. There is a long tradition, for instance, in lesbian oral history of charting the intricate details of class divisions within particular communities (see Nestle 1987; Davis and Kennedy 1989). As a reaction and response towards queer theory recently some writers have started to think through the class politics behind queer politics and activism. For instance Peter Cohen's (1997: 87) study of AIDS activism in New York shows that for many middle-class gay men in the United States AIDS represented a 'class dislocation'. As he notes: 'AIDS dislodged certain gay men from their tenuous position within the dominant classes by transforming unmarked individuals into members of a stigmatised group.'

Allan Berubé is another writer and activist concerned with the class basis of gay and queer politics. Writing about his experiences as an activist in the gay community in San Francisco, Berubé reflects on his class difference as it impacted on his ability to participate in the gay social world of the city: 'what I experienced most directly as a white gay man with little money and no college degree was how the gay community reproduced class hierarchies' (1996: 152). He argues that much of the scene reproduced wider material inequalities and he was unable to participate in it.

However, class is not just a matter of income; it is being increasingly defined through culture (by government rhetoric, media representations, marketing speak, economic discourse, and practice, such as consumer choice or safety behaviour: see Skeggs 2003), in which certain forms of culture become 'propertisable', that is, they can be converted into exchange-value which enables people to move through social space with entitlement and access to a wider range of areas than those without the requisite capital.[2] As Bourdieu (1979, 1986, 1987, 1989) notes, the way we move through social space with different types of capital (cultural, social, symbolic and economic) enables us to embody different volumes and compositions of these capitals.[3] Social class is made up of this embodied combination, lived and carried on the body, displayed through dispositions such as the access, entitlement and occupation of space. Class is always composited with other forms of capital and social positioning such as race, gender and sexuality which provide different values in the volume and composition; these amalgams can both enable and restrict access and entitlements but are always visible through dispositions.

For Ulf Hannerz (1996: 103) cosmopolitanism is a disposition, an attitude:

> A more genuine cosmopolitanism is first of all an orientation, a willingness to engage with the Other. It entails an intellectual and aesthetic openness toward divergent cultural experiences, a search for contrasts rather than uniformity. To become acquainted with more cultures is to turn into an aficionado, to view them as artworks. At the same time, however, cosmopolitanism can be a matter of competence, and competence of both a generalised and a more specialised kind. There is the aspect of a state of readiness, a persona; ability to make one's way into other cultures, through listening, looking, intuiting, and reflecting. And there is cultural competence in the stricter sense of the term, a built-up skill in manoeuvring more or less expertly with a particular system of meanings.

A more detailed articulation of Bourdieu's analysis of how class is displayed through cultural knowledge and disposition would be hard to find. To turn the intellectual gaze into a form of knowledge and competence for one's own enhancement is precisely how cosmopolitanism as a disposition is generated through the ability to move through the cultures of others, turning them into objects of distanced contemplation for oneself. Intellectual cosmopolitans produce themselves by travelling through the cultures of others, others who may not have the requisite capitals to return the gesture. Access to the world of the intellectual aficionado may be far more difficult to acquire. Chris Haylett (2001) argues that in contemporary British society white working-class culture is the abject constitutive limit by which middle-class multicultural citizenship is known and valorised. If we consider the way class has been configured within cosmopolitanism we can see that the white working class are often represented as being either the enemy of cosmopolitanism, or simply incapable of demonstrating the cosmopolitan virtues discussed by Bryan Turner (2000). In his essay on cosmopolitan virtues it is significant, but perhaps unsurprising, to discover that the working class are configured as the inhabitants of non-cosmopolitan space:

> Those sections of the population which are relatively immobile and located in traditional employment patterns (the working class, ethnic minorities and the under classes) may in fact continue to have hot loyalties and thick patterns of solidarity. In a world of mounting unemployment and ethnic tensions, the working class and the inhabitants of areas of rural depopulation may well be recruited to nationalist and reactionary parties. Their worldview, rather than being ironic, becomes associated with reactionary nationalism.
>
> (Turner 2000: 141–2)

So, predictably (as Haylett demonstrates in relation to New Labour rhetoric and the reactionary UK popular press, or as Ehrenreich (2001) reveals in the US press), the working class is associated with the reactionary politics of nationalism rather than the multicultural politics of cosmopolitanism. It has been evident throughout our discussion of the discourses of cosmopolitanism that the concept is highly contested. In terms of sexuality, gay men's relationship to cosmopolitanism has been transformed from Cold War suspicion and abjection, to post-Cold War symbols of globalised conspicuous consumption. In terms of class politics white working-class difference has become significant in valorising middle-class cultural identities. As we stated at the very start of this section, both sexuality and class are routinely and commonly seen as sources of anxiety, discomfort and unease. In terms of the spatial politics of cosmopolitanism, it is imperative that we recognise that cosmopolitanism operates through access to and knowledge of specific places and spaces. Cosmopolitanism offers the promise of footloose-ness, against the reactionary nature of place-based identities – a rejection of fixed identities. However, it is the case that specific places within the city are commonly seen or promoted and branded as displaying the virtues of cosmopolitanism. They are also seen as sharing these values with connected spaces across the globe.

Producing cosmopolitan space: Manchester's Gay Village

> The city becomes not so much a complex of dangerous and compelling spaces of promises and gratifications, but a series of packaged zones of enjoyment, managed by an alliance of urban planners, entrepreneurs, local politicians and quasi-governmental 'regeneration' agencies. But here, once more, urban inhabitants are required to play their part in these games of heritage, not only exploiting them commercially through all sorts of tourist dependent enterprises, but also promoting their own micro-cultures of bohemian, gay or alternative lifestyles, and making their own demands for the rerouting of traffic, the refurbishment of buildings, the mitigation of taxes and much more in the name of the unique qualities of pleasure offered by their particular habitat.
>
> (Rose 2000: 107)

One of the main reasons why we were originally drawn to the concept of cosmopolitanism was the numerous times the term was used to describe the Gay Village by those consuming and producing the space in our research. We read this as a concern to articulate the desirability of the space, whilst also acknowledging the negotiation of difference.

In the next section we argue that perhaps the term cosmopolitanism is used to avoid embarrassment about discussing the consumption of difference. In this section we argue that the branding of the space as cosmopolitan is part of a strategy to make the space less threatening, hence a more appealing and desirable space of consumption for a wider, straight community. The more threatening, less easily assimilated aspects of urban sexual dissidence are rendered invisible – and most specifically the sexual side of gay men's urban cultures are downplayed, with only certain aspects of gay male culture promoted. Therefore we examine the production side of space – and consider the ways in which the Gay Village has become part of what enables Manchester to be seen as a model of urban regeneration. This section then focuses on the place-marketing of space, and the political economic framework for the discussion of consumption that follows.

We now explore the space through which this argument will be developed, drawing on the research project conducted between 1998 and 2001 on 'Violence, Sexuality and Space'.[4] The aim of the project was to understand the production and sustainability of gay safe space. Lancaster – a much smaller city in the north-west of England, which has no commercial gay space – was used as a comparison, but will not be drawn upon here (see instead Moran and Skeggs 2003).[5]

During the 30 months of the research we interviewed 58 key informants, conducted two major surveys of nearly 1000 'gay space' users, ran 35 focus groups in total (with each group of gay men, lesbians and straight women) in the two different locations, plus two marginal case study groups with transsexuals. We analysed secondary data (economic, geographical and council reports) and media data (magazines, tourism data and TV and film productions of the space, e.g. the impact of the TV series *Queer as Folk* on the space), and conducted continual (participant) observation. The research constantly fed different elements of the distinct methods into each other. So observation, report and representation reading informed the questions asked of the key informants; their responses then framed the questions for the survey (being read with and against research literature and observation). The findings of the survey were then fed into the focus groups, whose challenges were then incorporated into the two Citizen's Inquiries we conducted with all research participants at the end of the research. The discussion following the Citizen's Inquiries fed back into the recommendations we have made (see www.les1.man.ac.uk/ sociology/vssrp). One of the benefits of a large interdisciplinary research team is that we were able to cover such a huge amount of different data (we all had different research skills); on the other hand, it meant we often disagreed, but these disagreements were productive, becoming part of the agenda-setting nature of the research, and researchers were encouraged to develop areas that they found of special interest.

Manchester's Gay Village emerged following a huge homophobic policing campaign staged by Chief Constable James Anderton, to become one of the most visible, compact and gentrified gay spaces in the UK (Whittle 1994; Quilley 1997). Anderton's moral campaign was significant to the present 'structure of feeling' of the Gay Village as it produced organised resistances, which were later institutionalised in the present Labour city council. Known gay spaces have existed in Manchester since the early 1900s but the difference is that they were not consolidated by compact visibility. The 'Gay Village' became more visible to the heterosexual community during the 1990s with the development of an innovative modernist architecturally designed bar with 30-foot glass windows, called Manto. The architectural design was a queer visual statement: 'We're here, we're queer – so get used to it.' It was a brick, glass and mortar refusal to hide any more, to remain underground and invisible.

Manto was developed on a space, which used to be the Worker's Reading Library, opposite the Rochdale Canal (the heart of the industrial revolution) and very close to three old traditional gay bars and two major cottages (public sex spaces of the George Michael kind). It is in close proximity to train and bus stations, just outside of the main shopping centre, next to Chinatown and very close to three large universities (which have a combined population of 62,000: not insignificant when only 450,000 live in the city centre = 14 per cent). The Gay Village offers at least 50 different opportunities for gay consumption of varying kinds. The surrounding streets have been developed with very expensive loft apartments, more bars and restaurants and in 1998 the first lesbian bar, Vanilla. Until the opening of Vanilla, lesbian space in the village was restricted to one room above a male leather bar on Canal Street that two years ago was turned into a 'bistro'. The lesbian club Follies closed in April 2001. It had a history as a significant space for working-class lesbians throughout the greater Manchester area.

Taylor *et al.* (1996) demonstrate how Manchester has a long tradition of cultural entrepreneurship, which Mellor (1997) traces as an influence on the 1980s economic regeneration route based on 'culture'.[6] This model has since been used by other cities. The village has been strategically incorporated into this 'culture-led' programme, and through a variety of local governmental factors (Quilley 1997, 1998) has become central to the marketing vision of Manchester: 'It was a very downtrodden part of the city centre and isn't now, it is now the most desirable part, it is the part that is projected across the world as the face of the new Manchester' (John, marketing director, Metz).

The aesthetically pleasing village is constantly represented as cosmopolitan/European in its publicity, and the comparison is often made to sunny European Barcelona rather than other UK cities: a rather surprising image for those who know the rest of this wet Victorian northern town. However, it can look good. Many of the bars in the village on the central drag, Canal Street, have outside seating and one bar, Metz, has a restaurant, bridge and boat on the canal. In spring and summer, when packed with promenading (or cruising) gay men, lesbians, tourists of all varieties and the smart and trendy clubbers, it could at moments be a vision of a sophisticated space. It was represented in the Channel Four television series *Queer as Folk* (first screened first in 1999, series 2 in February 2000) as a utopian hedonistic gay male space, with no violence, tolerant and cosmopolitan. Campaigning group 'Healthy Gay Manchester', who spatialise and mark their presence in the village through adornment with banners and flags, describe the village as the pride of Europe:

> Manchester's stature as a European gay Mecca continues to thrive as visitors flock from all over to sample its unique spirit. At the heart of this fast-growing, post-industrial city sits the eclectic cafe society of the Lesbian and Gay Village – a cosmopolitan showcase bursting with pride, and one of the queerest pieces of real estate Europe has to offer.
>
> (Healthy Gay Manchester 1998: 1)

While the development of Manchester's Gay Village may have been something new in Britain in the 1990s, we have also witnessed the rush to develop similar spaces within other cities such as Birmingham. The use of the term eclectic suggests a space that is authentic, rather than branded or corporatised, and hints at an embrace of difference. Manchester's promotion of its Gay Village must be seen in the wider context of place promotion under neo-liberalism. The promotion of such cultural quarters has become a feature of contemporary British urban policy (O'Connor and Wynne 1996; Bell and Jayne 2004). In the British context Manchester has been seen as at the vanguard of these policy initiatives based on the promotion of cultural entrepreneurialism and the establishment of public–private initiatives (Quilley 2002; Robson 2002). This policy sets Manchester in competition with other cities, on both a national and European scale. In this context Neil Brenner (1999: 433) argues that 'on sub-national spatial scales, interspatial competition has intensified among urban regions struggling to attract both capital investment and state subsidies'. Manchester has been very successful in making use of the Gay Village in its wider city promotion efforts. Rose (2000: 106) argues that transgression is increasingly becoming commodified and that alliances between city politicians and entrepreneurs have sought 'to reshape the real city according to this *image* of pleasure, not least in order to enter into the competitive market for urban tourism'. It is an image of imaginary cosmopolitanism. For instance, global gay tourism has been seen as an important component in Manchester's marketing policy: John from Metz, who was mentioned earlier, was part of a team that internationally promoted Manchester tourism, the village being central to this enterprise. And in an interview Joanna from *Marketing Manchester* speaks of the Gay Village as important for attracting investment and tourists to the city:

> it is obviously a unique selling point that Manchester has. It has taken time to get it accepted by some of the board in our organisation, however the fact that the council are seen to be gay friendly, the minister of tourism is, or it comes under his department, and the BTA [British Tourism Authority] have now decided to actively and openly promote the UK, it has gained credence and I've been able to suggest that this could be an official policy now and it is one of our target markets.

We therefore see how the Gay Village is used to market the city as cosmopolitan and user-friendly. We argue that the use of gay males in particular to mark out cosmopolitanism has depended on them remaining in the position of safe, usable other, but also on a significant proportion remaining still different: strange enough to be unthreatening but respectable.

We now examine how this space is consumed. In the first section of the chapter we argued that most definitions of cosmopolitanism contained a desire for a more authentic engagement with Others, an imaginary, distanced cosmopolitanism. We shall now focus on how this desire for difference works out in practice in proximity. We initially establish the inscribing processes of value, then move to an analysis of how straight women in particular use the village.

Knowing and negotiating urban difference

Earlier in the chapter we also argued that to be cosmopolitan one has to have access to a particular form of knowledge, to know how to appropriate and recognise valued differences and to generate authority from this knowing. Cosmopolitanism, we argued, is not just about movement through space but is an embodied subjectivity reliant on the requisite cultural capital for generating the requisite dispositions. The supposed European nature of the space is repeatedly mentioned in focus groups and interviews with users of the village. For instance, focus group participant Ruth:

> It's an area of Manchester people go there, lots of nice bars, it's becoming like a European city with that sort of nightlife and so all these people are – all these other bars are springing up around it because they are catching on to a good thing.
>
> (Ruth, Lesbian Focus Group no. 4)

Sandra, another participant in the same focus group, discussing the use of space in the village states:

> I like that kind of space . . . it's got that kind of continental feel, I think it's the only place in the whole of Manchester that has that European – you could be in Paris you could be in Barcelona.
>
> (Sandra, Lesbian Focus Group no. 4)

Later on in the focus group she adds: 'It's just got that European, *very advanced*, very now feel and for me they've got a good selection of soft drinks.'

These comments express a desire for identification with the more tolerant, open and liminal spaces associated with cities such as Amsterdam and Barcelona, demonstrating the success of attempts to brand the space as European and cosmopolitan. They also reflect a yearning for affluence and status associated with foreign travel, and aspirations of and glamour associated with all things European. In Manchester the very branding of the venues and residential developments that have opened in the major expansion of the late 1990s reflect a conscious attempt to capture and commodify the glamour associated with cities in the United States and Europe. For instance, new bars such as Tribeca, Metz and Prague, as well as new residential developments such as the Lexington, directly signal the attempt at branding. Not all attempts to brand the area as European have succeeded, however – witness the closure in 2000 of a Belgian-themed bar and brasserie in a prime location on Canal Street. Those who own businesses in the village market it as an integrated, mixed space which has one essential feature, 'gayness': 'Obviously the village is grown on the back of the fact that this is a lesbian and gay space. The reason that people come here is for the gayness of it, even if they are not gay' (John, marketing director, Metz).

The essentialising of gayness is a requirement for marketing the area: as our focus groups have consistently shown, it is the 'gayness' of the area that makes it European *and* unique, that gives it its cosmopolitan character which is used to draw people of other sexualities into the village. This is summed up by Julie, a straight woman from our focus group: 'Gay people just have this liberated – you can do whatever you like type' (Julie, Straight Women's Focus Group no. 2).

Gayness can be branded as user-friendly, making it into a quality that can be read back onto the space: it is not without irony that the largest and most successful of the clubs in the village is called Essential.[7] In this context Dereka Rushbrook (2002: 195) argues that: 'Gay urban spectacles attract tourists and investment; sexually deviant, dangerous rather than risqué, landscapes do not.' There is also an obvious paradox built into this, for if too many straight people enter the village its essential quality of authenticity, gayness, will be diluted.

Our research has shown consistently that the major threat all users of the village feel, which disrupts their use of the space, comes from 'hen parties' and 'scallies' – a term used to describe white working-class men in the north-west of England:

> If you walked in a bar and it was full of scallies like Dickens, maybe you'd think – a bit wrong you know whatever you might – it's where you feel comfortable isn't it. Yeah it's where you feel comfortable. You do get the vibes don't you? And you either stay and enjoy it or have a drink and oh right, I'm off.
>
> (Paula, Straight Women's Focus Group no. 4)

> I think that I would be more comfortable with a group of gay men that were drunk than I would be in a group of lads that were drunk just because they were more politically aware – I just don't like this a group of lads going out and getting drunk and 'whoaaa' that type of attitude, the same with a group of women as well and you don't get that in the gay village.
>
> (Cath, Straight Women's Focus Group no. 2)

This is not the process of gaining pleasure through the visual and distanced contemplation of difference, a disposition which Bourdieu (1986) defines as central to the production of the bourgeoisie, but of difference in proximity, of being close to unthreatening authentic difference.

In an ethnographic study of white working-class women, Skeggs (1997) notes how the direct expression and claim to 'be sophisticated' by the women represented a classed statement of positioning: only those positioned at a distance from sophistication would declare themselves so desirous of it (an insensitivity noted earlier by Litvak). Moreover, Litvak argues that part of the embarrassment of those who know to maintain a distance from a desire for sophistication is the squeamishness about the associated (homo)sexuality of sophistication, which, he argues, is read through femininity and historically associated with perversion or adulteration. Cosmopolitanism as a claim to authority, as a way of achieving unspoken sophistication, is always classed and sexed. Certain objects are chosen to enable cosmopolitanism to be achieved. These objects (or groups) have already been allocated cultural value and are used to bolster the cultural capitals of the user.

The cosmopolitan is produced through consuming difference, but only certain differences. The gay man is frequently positioned as necessary for that consumption as a signifier of difference, but this consumption of difference is only available to those who can (and want to) access different cultures. Gay men who want to be fixed (if only momentarily), recognised and known, to belong and stay in place for a time, can only be cosmopolitan by existing as a sign of difference for others. And even the most marginalised space users can access the 'essential' quality of user-friendly gayness to enable themselves to feel not only safe, but also European and cosmopolitan.

We have also seen that the term cosmopolitan can be used to avoid embarrassment. No one using the space wants to be seen as invading gay male space, or wishes to be accused of being uncomfortable in the presence of gay men, or of homophobia. Hence the resort to terms such as cosmopolitan and comfort to articulate ease within space. Thus, after Hannerz, the true cosmopolitan negotiates his [sic] way through different urban situations – in a gay space this equates to an ease of behaviour – either knowing how to dress to pass, or at least not to stick out, learning how to 'play' with difference in the negotiation of space.

Consuming difference – how straight women consume the village

> We use the village because we don't get hassle from straight men. So you could go to a bar with women friends, sit and talk about what you want to talk about and you wouldn't get that hassle.
>
> (Cath, Straight Women's Focus Group no. 4)

> You're literally ignored in that club, I've never felt so free in my life – I can just do basically things that I would never dream of doing in a straight club and it might be because . . . [Facilitator: it's liberating?] It is and it's really like I don't mind seeing myself as a person that just because there are straight men there that I feel I have to suit and I hate that but yet and maybe it is the drink that does help, but you just feel kind of – I remember the last time I went in there there was a group of us and I think we literally jumped up and down for about seven hours and after I thought what did we just do.
>
> (Caroline, Straight Women's Focus Group no. 2)

It is evident from these quotes that the straight women in our study use the space because it is gay. They also use it because of a shared cultural capital which is committed to glamour, hedonism and anti-pretentiousness and because it used to be the most trendy place in Manchester (see Skeggs 2001). The manager of one of the most trendy, young, glamorous bars (used in *Queer as Folk*) shows how it is the link between trendiness and sexuality that is at stake for consumption:

> [Interviewer: Is Manto's not the right place for straight women?] Trendy women, yeah. Trendy, but the normal everyday Sharon and Tracy I would say 'No'.
>
> (Steph, bar manager, Manto's)

Here we can see that a particular sort of working-class woman is not welcome. As John, marketing director of another trendy bar, Metz notes:

> Well I would say Manchester, probably more working-class people feel comfortable; no that's not true, in this bar they would and the New Union they probably would; maybe Via Fossa's geared more to the middle classes, Manto's, Metz, you know. I think, depending on your class will depend on which bar you go to and which bar you feel comfortable in.

For some of the straight women in the focus groups, the village represents a 'comfortable' space where they can belong, a place of safety, but, as Ruth Holliday (1999) argues, notions of comfort can be used to produce essentialist ideas (see also Moran and Skeggs 2003). While the village constitutes a comfortable space for some of these women, their sense of comfort and ease in the village may produce discomfort and dis-identifications among other users of the space.

Straight white working-class men are the group identified as being the main threat to the sophisticated image of the village. In the straight women focus group, one woman complains that the village is becoming a less safe space because of the influx of groups of young straight working-class men.

> It's this other element that both the straight community that go out with the gay community both really feel quite strongly about, yet I don't go into the village any more because I just don't like the sort of the element that I think is sort of sometimes, not all the time you know, makes a really negative, horrible impact which makes people aggressive and therefore there's this backlash of you know that gay village for gay people you know.
>
> (Zoe, Straight Women's Focus Group no. 4)

> It's that bad egg element that really homophobic element that still likes coming to the village when they are really homophobic and then what are they getting out of it. That element stopped me coming into the village.
>
> (Zoe, Straight Women's Focus Group no. 4)

Straight women complain that the influx of straight men makes the space less attractive and less safe for them. There is a fundamental contradiction here between the women's desire and entitlement to use the village, and their concern that the village is becoming 'too straight'.

> I think I do have a right to be there as a straight woman in a gay space. I feel more comfortable when I am with gay friends and if I'm walking down to Canal Street with my partner we won't hold hands like we do normally, that's a bit off, I don't know what that is really, you can analyse that in different ways, but to me it's like a little bit of respect for that gay space and sexuality.
>
> (Caroline, Straight Women's Focus Group no. 2)

Caroline is keen to assert her respect for gay users of the village. It is this knowledge – of how to conduct oneself in gay space – that she uses to draw a distinction between herself and other straight women. The key difference for her is between those groups of women who know how to behave in gay space, and those who clearly do not know any better. She thereby constitutes herself as cosmopolitan, as possessing this knowledge about how to behave in 'gay space', against the unsophisticated, provincial women who have no respect:

> People could say it's becoming straighter and I personally would say there is nothing wrong with that as long as there is this idea that this is the gay village and you respect everybody here for what they want to do, where they want to go and how they are dressed. And I think that is why for me it is such a place that is welcoming because people seem to be accepting of people. But if it is becoming straighter and that means that the straighter people are pushing the gay people out, then I think that's bad. *Something should be done about that but I don't think there is anything wrong with it becoming a straight and gay place that would be worth promoting I would have thought.* As long as that respect is there but a lot of people do come into the village, like little gangs of girls and they do treat it a bit like a freak show . . . a lot of them don't have that respect and I think that's when the trouble starts to set in.
>
> (Caroline, Straight Women's Focus Group no. 2, emphasis added)

There are clear class differences between these groups based on education and knowledge about appropriate behaviour – cosmopolitanism as knowledge. 'Little gangs of girls' combines the rough with the infantile: 'But it is all to do with education you know, we all mix with all sorts of people. Those types of people just don't' (Caroline, Straight Women's Focus Group no. 2). Uneducated working-class people who have not mixed with gay men who do not know how to behave in the Gay Village are not cosmopolitans, but are spatially located as provincial, outside of the urban. Caroline goes on to argue that cosmopolitan knowledge is a matter of place:

> Everybody gets to meet gay men even if they would be, you know what a great guy, no attitude no homophobia, it's just how did I get to meet gay men, the core group in Macclesfield. I'm sure if I'd lived where I used to live in Knutsford I wouldn't have met a gay man, do you know what I mean? It's purely whatever, do you know what I mean, I could have ended up being up, I don't know it's just . . . It's the little towns, like it's a bit of socialisation in terms of your education . . . and some of that will be the views and attitudes that you've got to what you've grown up, you're coming to look at something and then it's like experiences as well as when you get older.
>
> (Caroline, Straight Women's Focus Group no. 2)

While the white straight working-class women in the focus group celebrate and position themselves as imagined cosmopolitans, Julie argues that there are few spaces for black women, and that she does not feel much sense of community in the village:

> My sense of community is that now because the Black community, there is

> no Black community in the club world in Manchester, so wherever I go there is always that feeling of isolation to a point because there's no-one that looks – they are playing my music or the music of Black culture but there is no-one around that looks like me or very few, so that's my next thing that sense of community.
>
> (Julie, Straight Women's Focus Group no. 4)

And Sue:

> It got to that point where I started thinking because as well, when I used to go out, especially at Cruise, I never got let in and I questioned the bouncers on the door and I said it was a race issue and that, and then they promptly let me in. But a lot of times I couldn't even get into places and it wasn't because I was a straight woman because I have seen straight women just walk in so I – sometimes I feel my presence on two levels, as the only black person in there, or you know, as a straight woman in there.
>
> (Sue, Straight Women's Focus Group no. 3)

As Schein (1999) details, in other spaces race works to enable imagined cosmopolitanism, but here in a place where the essential authentic branding ingredient is sexuality, race has no place. It disrupts the homogeneity of the user-friendliness. So, whilst the white straight working-class women position themselves as the ideal users of the space through their knowledge of how to perform cosmopolitanism and respect, others cannot even make that gesture. They are excluded. It may be the combination of gender and race that disrupts the reading of sexuality as male and white.

Conclusion: exclusions from cosmopolitan space

In this chapter we have seen that the label cosmopolitan is repeatedly used to brand Manchester's Gay Village by those who produce the space. The label is used to promote the neighbourhood in a non-threatening way, to broaden its appeal to consumers regardless of sexuality, as a way of bringing in customers and money. The 'gayness' of the village is promoted to the wider community as a non-threatening authentic commodity. In this sense the label cosmopolitan is all embracing. To be sophisticated and cosmopolitan means to know how to occupy particular spaces in the city. Within Manchester this means to participate in the consumer spaces of the city's Gay Village. What we see is a complex cartography of competing definitions of sophistication and cosmopolitanism.

Our analysis suggests that we are at the stage of late capitalism identified by Žižek, in which multicultures can be made corporate through their 'essential' difference. They are partially incorporated for profit, enabling divisions to be drawn between the profitable and the non-profitable. This suggests that it may be marketability rather than sexuality that is the issue for marginalised groups for the next millennium. Are our differences worth knowing, having, experiencing? The partial destruction of the heterosexual family, and the range of living relationships now being made possible, suggests that the 'danger within' identified in the homophobic Thatcher years may have paradoxically been eclipsed by the demands of the market. It is ironic that campaigns against homophobia, resulting in the creation of 'safe gay space', have themselves become subject to market forces.

However, this process of incorporation through imaginary cosmopolitanism is not entirely straightforward. There are cosmopolitan winners and cosmopolitan losers. Gender, race and class have intervened in the commodification of gay space. Moreover, whether the use of this space by lesbians and gays makes them complicit with capitalism is still open to debate, for who isn't? It certainly does not guarantee that lesbians and gays take an uncritical stand towards the territorialisation of gay space. In fact, our research suggests that because they struggled most and have the greatest investment in the space, they are its most critical users. Yet we can still see how gay men, lesbians and transgendered people become objects for the fetishisation of difference.

Underlying intellectual and capital appropriation is the central mechanism of distinction that distinguishes which forms of gay and lesbian cosmopolitanism are worth knowing about, worth exploiting, worth excluding. Some of the straight women in the research travel through the spaces with subcultural value, accumulating knowledge and enhancing their own reservoir of cultural capital, yet others cannot do the same, as Julie and Sue demonstrate. In terms of the spatial politics of cosmopolitanism, Manchester's Gay Village was originally promoted as being a special, 'Other authentic' place – different from the mainstream of consumption spaces within Manchester. Yet with gayness being diluted through use, simultaneously reclaimed, and made user-unfriendly, and with the development of copycat, but not gay spaces elsewhere (Deansgate Locks, with the attraction of Manchester United footballers), the cosmopolitan nature of the space can be seen as just another attempt at banal consumption by branding. For a time an imaginary cosmopolitanism, premised on an 'essential' sexuality (figured primarily as male and white), was on offer for the fickle straight consumer that is now elsewhere, but that was not without its social struggles, as the straight women's usage showed. Fulfilling the logic of late capitalism identified by Žižek differences are incorporated, enhancing the cultural capital of the already privileged, establishing others as the constitutive limit. But as the search for distinction escalates (by both marketers and the subcultural accruers), gay men and lesbians can more safely reinhabit the space in which for a short space of time they became the motif of urban regeneration, the soul of the new consumer.

Notes

1 See Skeggs (2003) for the class-based formation of the concept and experience of 'self'.
2 Propertising, a term taken from legal studies (e.g. Coombe 1993; Radin 1993; Davies 1994, 1998, 1999), establishes a relationship between property and personality. As a concept it enables an investigation into how some people make investments in their cultural characteristics, which can then be used to realise a value in areas (such as the economic) protected by legal property rights. For instance, being heterosexual enables one to enter parts of the labour market that would be closed off to queers. Inhabiting the position of heterosexual enables people to realise benefits (from welfare and tax systems of the state, insurance systems, property ownership of private capital). It also provides legally given rights to privacy over one's sexuality, which are not available to male queers, who are constantly criminalised and have no access to the private (Berlant and Freeman 1993). We need to know which cultural characteristics have a legal privilege and entitlement that enable propriety to be converted into property, legitimated and authorised.
3 Bourdieu offers contradictory explanations as to whether some social positions constitute fields or cultural capital, so whereas he sees class and race as a dynamic relation of domination that pervade all fields, he sees femininity as capital specific, whilst masculinity is an across-field form of domination (see McCall 1992; Lovell 2000).
4 The project team comprised Leslie Moran, Beverley Skeggs, Karen Corteen, Paul Tyrer and Lewis Turner. ESRC Award Number L133 25 103.
5 Focusing on Manchester, we acknowledge criticisms of the urban bias within work on the geography of sexuality and the need for more research on neglected areas of study such as the rural (Bell and Valentine 1995; Binnie and Valentine 1999; Elder 2000), which as yet in the UK has not been directly marketed for global gay consumption.
6 For those who have not yet spotted the connection, the culture-based nature of new class analysis is not separate from, but rather an attempt to understand, new forms of value that emerge from the whole raft of 'cultural policy development' that is taking place across a range of sites alongside the urban.
7 It is owned by Nigel Martin Smith, who as manager of Take That was very successful at marketing gayness disguised as heterosexuality to a mixed audience.

References

Altman, D. (1996) 'Rupture or continuity? The internationalization of gay identities', *Social Text*, 48: 77–94.

Altman, D. (1997) 'Global gaze/global gays', *GLQ: A Journal of Lesbian and Gay Studies*, 3: 417–36.

Altman, D. (2001) *Global Sex*, Chicago: Chicago University Press.

Beck, U. (2000) 'The cosmopolitan perspective: sociology of the second age of modernity', *British Journal of Sociology*, 51: 79–105.

Bell, D. and Binnie, J. (2000) *The Sexual Citizen: Queer Politics and Beyond*, Cambridge: Polity Press.

Bell, D. and Binnie, J. (2002) 'Sexual citizenship: marriage, the market and the military', in D. Richardson and S. Seidman (eds), *Handbook of Lesbian and Gay Studies*, London: Sage.

Bell, D. and Jayne, M. (eds) (2004) *City of Quarters: Urban Villages in the Contemporary City*, Aldershot: Ashgate.

Bell, D. and Valentine, G. (1995) 'Queer country: rural lesbian and gay lives', *Journal of Rural Studies*, 11: 113–22.

Berlant, L. and Freeman, E. (1993) 'Queer nationality', in M. Warner (ed.), *Fear of a Queer Planet: Queer Politics and Social Theory*, Minneapolis, MN: University of Minnesota Press.

Berubé, A. (1996) 'Intellectual desire', *GLQ: A Journal of Lesbian and Gay Studies*, 3: 139–57.

Binnie, J. (2004) *The Globalization of Sexuality*, London: Sage.

Binnie, J. and Valentine, G. (1999) 'Geographies of sexuality – a review of progress', *Progress in Human Geography*, 23: 175–87.

Bourdieu, P. (1979) 'Symbolic power', *Critique of Anthropology*, 4: 77–85.

Bourdieu, P. (1986) *Distinction: A Social Critique of the Judgement of Taste*, London: Routledge.

Bourdieu, P. (1987) 'What makes a social class? On the theoretical and practical existence of groups', *Berkeley Journal of Sociology*, 32: 1–17.

Bourdieu, P. (1989) 'Social space and symbolic power', *Sociological Theory*, 7: 14–25.

Brennan, T. (1997) *At Home in the World: Cosmopolitanism Now*, Cambridge, MA: Harvard University Press.

Brenner, N. (1999) 'Globalisation as reterritorialisation: the re-scaling of urban governance in the European Union', *Urban Studies*, 36: 431–51.

Cheah, P. (1998) 'Introduction Part II: The cosmopolitical – today', in P. Cheah and B. Robbins (eds), *Cosmopolitics: Thinking and Feeling Beyond the Nation*, Minneapolis, MN: University of Minnesota Press.

Clarke, D. (1993) 'Commodity lesbianism', in H. Abelove, M. A. Barale and D. M. Halperin (eds), *The Lesbian and Gay Studies Reader*, London: Routledge.

Cohen, P. (1997) '"All they needed": AIDS, consumption, and the politics of class', *Journal of the History of Sexuality*, 8: 86–115.

Coombe, R. (1993) 'Publicity rights and political aspiration: mass culture, gender identity and democracy', *New England Law Review*, 26: 1211–80.

Davies, M. (1994) 'Feminist appropriations: law, property and personality', *Social and Legal Studies*, 3: 365–91.

Davies, M. (1998) 'The proper: discourses of purity', *Law and Critique*, IX: 147–73.

Davies, M. (1999) 'Queer property, queer persons: self-ownership and beyond', *Social and Legal Studies*, 8: 327–52.

Davis, M. and Kennedy, E. (1986) 'Oral history and the study of sexuality in the lesbian community: Buffalo, New York 1940–1960', *Feminist Studies*, 12: 7–25.

Delanty, G. (1999) 'Self, other and world: discourses of nationalism and cosmopolitanism', *Cultural Values*, 3: 365–75.

Ehrenreich, B. (2001) *Nickel and Dimed: On Not Getting By in America*, New York: Henry Holt.

Elder, G. (2000) 'Review of F. Browning "A Queer Geography: Journeys Toward a Sexual Self" and W. Fellows "Farm Boys: Lives of Gay Men from the Rural Midwest"', *Professional Geographer*, 52: 155–61.

Evans, D. (1993) *Sexual Citizenship: The Material Construction of Sexualities*, London: Routledge.

Florida, R. (2002) *The Rise of the Creative Class: and How it's Transforming Work, Leisure, Community and Everyday Life*, New York: Basic Books.

Frank, T. (1997) *The Conquest of Cool: Business Culture, Counterculture and the Rise of Hip Consumerism*, Chicago, IL: University of Chicago Press.

Frank, T. (2001) *One Market Under God: Extreme Capitalism, Market Populism, and the End of Economic Development*, New York: Doubleday.

Hannerz, U. (1996) *Transnational Connections: Culture, People, Places*, London: Routledge.

Haylett, C. (2001) 'Illegitimate subjects?: abject whites, neoliberal modernisation, and middle-class multiculturalism', *Environment and Planning D: Society and Space*, 19: 351–70.

Healthy Gay Manchester (1998) *Healthy Gay Manchester's Guide to Lesbian and Gay Greater Manchester*. Manchester: Healthy Gay Manchester.

Holliday, R. (1999) 'The comfort of identity', *Sexualities*, 2: 475–91.

Jackson, P. (1999) 'An explosion of Thai identities: peripheral genders and the limits of queer theory', paper presented at the IAASCS Second International Conference, Manchester Metropolitan University, July.

Lacan, J. (1977) *Écrits: A Selection*, London: Tavistock.

Litvak, J. (1997) *Strange Gourmets: Sophistication, Theory and the Novel*, Durham, NC: Duke University Press.

Lovell, T. (2000) 'Thinking feminism with and against Bourdieu', *Feminist Theory*, 1: 11–32.

Markwell, K. (2002) 'Mardi gras tourism and the construction of Sydney as an international gay and lesbian city', *GLQ: A Journal of Lesbian and Gay Studies*, 8: 81–99.

Massey, D. (1993) 'Power-geometry and a progressive sense of place', in J. Bird, B. Curtis, T. Putnam, G. Robertson and L. Tickner (eds), *Mapping the Futures: Local Cultures, Global Change*, London: Routledge.

McCall, L. (1992) 'Does gender fit? Bourdieu, feminism and concepts of social order', *Theory and Society*, 21: 837–67.

Mellor, R. (1997) 'Cool times for a changing city', in N. Jewson and S. McGregor (eds), *Transforming Cities*, London: Routledge.

Moran, L. and Skeggs, B. (2001) 'Property and propriety: fear and safety in gay space', *Social and Cultural Geography*, 2: 407–20.

Moran, L. and Skeggs, B. (2003) *Sexuality and the Politics of Violence and Safety*, London: Routledge.

Mort, F. (1998) 'Cityscapes: consumption, masculinities, and the mapping of London since 1950', *Urban Studies*, 35: 889–907.

Nestle, J. (1987) *A Restricted Country*, Ithaca, NY: Firebrand Press.

O'Connor, J. and Wynne, D. (eds) (1996) *From the Margins to the Centre: Cultural Production and Consumption in the Post-Industrial City*, Aldershot: Ashgate.

Quilley, S. (1997) ‘Constructing Manchester’s new urban village: gay space and the entrepreneurial city’, in G. Ingram, A. Bouthilette and Y. Retter (eds), *Queers in Space*, Seattle: Bay Press.

Quilley, S. (1998) ‘Manchester first: from municipal socialism to the entrepreneurial city’, *International Journal of Urban and Regional Research*, 24: 601–15.

Quilley, S. (2002) ‘Entrepreneurial turns: municipal socialism and after’, in J. Peck and K. Ward (eds), *City of Revolution: Restructuring Manchester*, Manchester: Manchester University Press.

Radin, M. J. (1993) *Reinterpreting Property*, Chicago, IL: University of Chicago Press.

Robbins, B. (1998) ‘Introduction part 1: actually existing cosmopolitanism’, in B. Robbins and P. Cheah (eds), *Cosmopolitics: Thinking and Feeling Beyond the Nation*, Minneapolis, MN: University of Minnesota Press.

Robbins, B. (2001) ‘The village of the liberal managerial class’, in V. Dharwadker (ed.), *Cosmopolitan Geographies: New Locations in Literature and Culture*, London: Routledge.

Robson, B. (2002) ‘Mancunian ways: the politics of regeneration’, in J. Peck and K. Ward (eds), *City of Revolution: Restructuring Manchester*, Manchester: Manchester University Press.

Rose, N. (2000) ‘Governing cities, governing citizens’, in E. Isin (ed.), *Democracy, Citizenship and the Global City*, London: Routledge.

Rushbrook, D. (2002) ‘Cities, queer space, and the cosmopolitan tourist’, *GLQ: A Journal of Lesbian and Gay Studies*, 8: 183–206.

Schein, L. (1999) ‘Of cargo and satellites: imagined cosmopolitanism’, *Postcolonial Studies*, 2: 345–75.

Skeggs, B. (1997) *Formations of Class and Gender: Becoming Respectable*, London: Sage.

Skeggs, B. (2001) ‘The toilet paper: femininity, class and mis-recognition’, *Women’s Studies International Forum*, 24 (3/4): 295–307.

Skeggs, B. (2003) *Class, Self, Culture*, London: Routledge.

Sklair, L. (2001) *The Transnational Capitalist Class*, Oxford: Blackwell.

Smith, A. M. (1994) *New Right Discourse on Race and Sexuality: Britain, 1968–1990*, Cambridge: Cambridge University Press.

Smith, M. P. (2001) *Transnational Urbanism: Locating Globalization*, Oxford: Blackwell.

Stychin, C. (1995) *Law’s Desire: Sexuality and the Limits of Justice*, London: Routledge.

Taylor, I., Evans, K. and Fraser, P. (eds) (1996) *A Tale of Two Cities: Global Change, Local Feeling and Everyday Life in the North of England*, London: Routledge.

Tsang, D. (1996) ‘Notes on queer ’n’ Asian virtual sex’, in R. Leong (ed.), *Asian American Sexualities: Dimensions of the Gay and Lesbian Experience*, London: Routledge.

Turner, B. (2000) ‘Cosmopolitan virtue: loyalty and the city’, in E. Isin (ed.), *Democracy, Citizenship and the Global City*, London: Routledge.

Wakeford, N. (2002) ‘New technologies and “cyber-queer” research’, in D. Richardson and S. Seidman (eds), *Handbook of Lesbian and Gay Studies*, London: Sage.

Werbner, P. (1999) ‘Global pathways: working class cosmopolitans and the creation of transnational ethnic worlds’, *Social Anthropology*, 7: 17–35.

Whittle, S. (1994) 'Consuming differences: the collaboration of the gay body with the cultural state', in S. Whittle (ed.), *The Margins of the City: Gay Men's Urban Lives*, Aldershot: Ashgate.

Žižek, S. (1997) 'Multiculturalism, or, the cultural logic of multinational capitalism', *New Left Review*, 225: 28–52.

Conclusion: the paradoxes of cosmopolitan urbanism

Jon Binnie, Julian Holloway, Steve Millington and Craig Young

It has been the intention of this collection to understand and critically explore the ways in which the term cosmopolitanism is deployed and mobilised in a variety of different urban contexts. With regard to this aim, the contributors to this book have sought to ground cosmopolitanism in the development and processes of the urban. What these different case studies most tellingly reveal is the multiple characteristics of this term as it is differently deployed in practices and discourses of the contemporary city. Our contributors have thus highlighted the multiple, contested and often contradictory ways in which cosmopolitanism has been applied to the city, in both academic and non-academic contexts. Moreover, the multiple contexts of cosmopolitanism's deployment reveal a number of critical issues that are central to any understanding of the term. So, whether it be in the literature of the political and social sciences, or its application in (non-academic) policy-related networks, or even its vernacular or 'lay' usage, visions and practices of cosmopolitanism are far from unproblematic. In this conclusion, therefore, we seek to review these issues and thereby generate a critical research agenda for further studies of cosmopolitan cities. We finish with some speculations, informed by the different case studies discussed herein, as to the future of the so-called cosmopolis.

The critical issues raised by the chapters in this volume trace the limits of cosmopolitan urbanism. Thus, cosmopolitan cities generate, and are often produced through, a whole series of thresholds and checks. Put differently, there is a series of breaks which map the contours of the cosmopolis. For example, the chapter by Haylett illustrates one limit to the realisation of the cosmopolitan city in the form of class relations. Arguably it is often the case that the production of cosmopolitanism and cosmopolitan spaces are realised within and reproduce asymmetrical class relations. The role of class in the production of cosmopolitan spaces and cities exposes the limits of the term itself. As Haylett shows, the deployment of cosmopolitanism involves the annihilation of class difference despite the concept's emphasis on valuing difference. Where then does class difference fit into the cosmopolis? Furthermore, Haylett highlights the way in which the production of cosmopolitanism is bound up with forms of state coercion. Thus the state, in its attempt to produce cosmopolitan spaces, is achieved through an 'enforced reasonableness' or a compulsion to be a cosmopolitan citizen. This enforcement involves state policies which seek active engagement with difference. In addition to the limit of cosmopolitanism engendered through class relations, a programme of what can be deemed cosmopolitan active citizenship puts certain brakes on the cosmopolis. Indeed, these programmes of cosmopolitan active citizenship are illustrated in another context in the chapter by Chan. Here the presence of ethnic minorities in the UK city of Birmingham is deemed an asset in processes of regeneration. In particular, ethnic entrepreneurialism is seen as beneficial in terms of the generation of an entrepreneurial urbanism and regional economy. Therefore, as Chan argues, cosmopolitanism is not produced here as an unconditional hospitality, but rather an invitation that comes with certain conditions and duties (see Iveson this volume; Barnett 2005; Chan 2005; Derrida 2000, 2001). This enforcement of empowered cosmopolitan citizenry is in stark contrast to Germain and Radice's view of a more 'organic' less planned and consensually built cosmopolitan city in their case study of Montreal. As such, a 'fuzzy' cosmopolitanism might generate more productive outcomes rather than state-led versions of cosmopolitanism. Similarly, Bodaar reveals how the designated non-cosmopolitan spaces of the city might be spaces where cosmopolitanism closer to its idealised version might be being performed, practised and developed. In effect, the desire for key urban policymakers and actors to celebrate, plan and develop a city's cosmopolitan status might act to set a series of limits on what the cosmopolis might be or indeed become, and how urban residents are meant to act and behave.

These limits of cosmopolitan urbanism are echoed through and are supplemented by the tensions inherent to the term itself and its operationalisation. For example, Latham's chapter reveals that despite the supposed universalism of cosmopolitanism, national imaginaries and histories are central to the realisation of cosmopolitan geographies. Thus cosmopolitan spaces of cities such as Auckland are realised in the context of national histories and geographies of tensions and intolerance. We must therefore be careful to acknowledge that spatial scale is a significant concern when exploring the cosmopolitan urban. In other words, where and at what scale does cosmopolitanism happen? In Bridge's chapter it is precisely at the local scale that cosmopolitan cities are being produced. Moreover, this makes further problematic the distinction made between the open and tolerant cosmopolitan and the bounded and insular local, which we explored in the introduction to this book. Here cosmopolitanism is not about transversality, but rather local spaces of realisation. Perhaps, then, instead of looking to the global scale for the realisation of cosmopolitanism, we might need to investigate the local and everyday spaces of cities to see how and in what form cosmopolitan urbanities are produced. Indeed, local, everyday and mundane practices and habits are central to the preconditions of cosmopolitan cities that Sandercock's vision of interculturalism are focused upon: it is at the local scale and the spaces of the everyday that practices of cosmopolitanism and planned or policy-driven versions of cosmopolitanism might productively intersect, moving cities beyond the limits and tensions that cosmopolitanism inevitably, it seems, produces. It might be, therefore, that it is through planning and promoting a series of preconditions performed through daily habits, as Sandercock argues, where new forms of cosmopolitan cities are developed. Perhaps it is through a version of civic welfarism as promoted by Haylett, as an alternative to cosmopolitanism, which might offer other useful ways of resolving some of the tensions and limits of cosmopolitan cities. To a degree in this collection we have focused upon the spaces of consumption to understand how cosmopolitan cities are being worked out. Perhaps, then, we need to follow Amin's (2002) lead and study other micro-publics, such as the workplace, hospitals and schools as a future direction to understanding cosmopolitan urbanism, as well as keeping a critical eye on the local policy networks that many contributors have shown are central to the cosmopolitan city.

One of the key limits to the cosmopolitan city highlighted in this volume arises out of the branding and marketing of certain urban spaces as cosmopolitan. This process, which centres on the production of certain enclaves deemed somehow cosmopolitan, produces its own exclusions. As Binnie and Skeggs reveal, the attempts to promote the Gay Village in Manchester as part of a cosmopolitan vision for the city only acts to pathologise certain forms and practices of sexuality. Paradoxically, then, it might be that those spaces not marketed as cosmopolitanism are in their mundane and everyday production more cosmopolitan. As Brown shows, more mundane spaces such as public toilets are sites where a cosmopolitan tolerance is performed away from the sights of market forces and the entrepreneurial state. Therefore, the production of quarters and enclaves in cities that are marketed as part of a cosmopolitan urbanism might actually be detrimental to a city where difference is tolerated, understood and valued beyond the contribution it can make to a city's economy and entrepreneurialism. Perhaps it is in the cracks of the entrepreneurial and neo-liberal veneer often applied to cosmopolitanism that something closer to a more productive engagement with and encounter between differences is being performed. Again, it might be in the spaces of the everyday that cosmopolitanism is being performed without the regulation of the state: something as mundane as buying goods from grocery stores as Germain and Radice illustrate or, as Brown discusses, the practice of non-heterosexual couples holding hands might be where and how tolerance and understanding is being realised and sustained. Furthermore, it might be that goods and objects are central to a cosmopolitan urbanism: as Latham shows, there are potentially no limits on cosmopolitanism when engendered through the circulation of materialities.

The inherent tensions of cosmopolitanism and the limits set by the deployment of this vision reveal that cosmopolitan urbanism is a paradoxical phenomenon (Robbins 2002, 2003; Chandler 2003). Moreover, these paradoxes are many. To finish this conclusion, then, we wish to set out these paradoxes and explore their consequences for how we might think and practise the cosmopolitan city. The first of these paradoxes revolves around the issue of spatial scale. Thus cosmopolitanism is often discussed in terms that place a priority on the global and transnational forms of citizenry and belonging. Yet, as we have seen here, the national is not something that can be simply bypassed in the realisation of this valorised worldliness and global outlook. In particular, we must not ignore how cosmopolitan always intersects in often contradictory and conflictual ways with national policies and geopolitical decision making. We would thus argue that it is far from being the case that cosmopolitanism points towards the eradication of national difference and the relevance of the scale of the nation in the formation and contestation of urban difference. For example, it seems impossible to discuss relations between different ethnic groups without developing a critical awareness of how national governments shape and direct immigration policies. These national policies and directives are always shaping how difference is encountered, understood and made sense of in the urban realm. Put differently, cosmopolitanism, despite its declared global universalism, is always nationally differentiated.

A further and second paradox of cosmopolitan urbanism that these chapters have in various ways exposed involves class difference. Thus it seems that difference organised around class-based identities has the potential to undermine cosmopolitan visions of tolerance and productive encounters with alterity. There are two ways in which this is manifested in the city. First, the rise of the new middle classes means that issues of social and cultural capital need to be brought further to the fore in discussions of what it is to be cosmopolitan. Too often, then, it is the display and deployment of cultural and social capital in the generation of classed identities which destabilises cosmopolitanism. The construction and sustenance of class asymmetries through one's ability to make sense of or approach difference in an appropriate manner sit somewhat uncomfortably with visions of cosmopolitan urbanism. If cosmopolitanism is about 'being at home in the world' (Brennan 1997), it is clear that there are social and economic class constraints that mitigate against the production of a cosmopolitan subjectivity. Moreover, and second, certain class differences very rarely (if at all) figure in the vision of culturally and socially diverse urbanities. Paradoxically, then, there has been little attempt to seize upon class diversity in the race to win the seemingly valuable status of 'cosmopolitan city'. In fact the opposite is more often the case: the vision of cosmopolitan urbanism is bound up with a denial or pathologisation of certain class fractions. Indeed, when class intersects with other cultural formations such as ethnicity or sexuality this process seems only to be further compounded.

A third paradox of cosmopolitan urbanism develops from the limits placed on encountering difference produced when such encounters are driven by market forces and policy directives. Is cosmopolitan consumption therefore complicit with the logic of late capitalism, critiqued by Žižek (1997), which promotes a safe form of multiculturalism? Thus celebrating, promoting and marketing the cosmopolitan city often involves the production of enclaves. Arguably this demarcated and bounded alterity, produced through planning, economic and policy networks, not only pathologises difference which exceeds the boundaries of legitimacy, but also directs the encounter with difference. In other words, we would argue that a prescribed encounter with difference is a regressive step. The colonisation of the encounter with difference, through, for example, the commodification of alterity, is a normative project that paradoxically undermines the progressive potential of cosmopolitan cities. Furthermore, this determination of the encounter occurs in the context of models of citizenship. Thus encountering difference is increasingly shaped in the context of the demands placed on individuals and communities that in turn generate certain rights. The most pressing question here is, who gets to define and demarcate the cosmopolitan citizen and thus how they interact or engage with cultural diversity?

Moreover, at the heart of this colonisation of the encounter, and the forms of citizenry which both produce and are produced through it, lie everyday practice and the mundane space of cities. Thus attempting to prescribe how and in what ways urban inhabitants interact and engage with one another involves attempts to control the realm of quotidian practice. Paradoxically, once more, it is precisely in the everyday spaces of the city that a progressive form of cosmopolitan practice might be generated. However, the mundane and banal spatialities and temporalities of everyday life are where our attention must be directed. In doing so we can begin to further understand how both progressive and regressive forms of cosmopolitanism are being realised and interact. The paradoxical and simultaneous potentiality of the everyday, as both a space and time of colonisation or progressive cultural politics, is possibly *the* realm for a future critical research agenda of cosmopolitan urbanism. Thus the banal and mundane spaces of encountering difference (whether those spaces include the street, shop, home or workplace, and however difference is manifested) need to take a higher profile in our research agendas rather than sustaining an abstract theorising of cosmopolitanism that often seems detached from the everyday. If this is the case, then it is crucial that as social scientists we are constantly aware that our visions of what counts as a progressive cosmopolitan urbanism does not itself result in another colonisation of the encounter. Put differently, wresting cosmopolitanism from market forces and regressive forms of policy should not mean replacing it with prescribed cosmopolitanism with its own set of limits. Two potential instances of this come to mind. First, we must be careful here not to reify or even romanticise the everyday as privileged space and time of progressive cultural politics and interculturalism. Second, in any critical project of cosmopolitanism we should be constantly aware of deeming certain engagements with difference as more or less 'authentic' or 'inauthentic'. Although difficult in some sense to avoid, critical analyses that simply reproduce such a duality arguably sustain the very exclusions that a paradoxical cosmopolitan urbanism seems to produce. Perhaps, then, instead of trying to operationalise the concept of cosmopolitanism, and thus seek out solutions which may reproduce the asymmetries or exclusions we seek to move away from, we need new forms of practice that generate their own concepts and solutions beyond prescription. Thus the performative potentiality of the urban encounter might be where forms of progressive cosmopolitan practice are generated.

Furthermore, in reviewing both the chapters in this book and the debates elsewhere in the social scientific literature, we have been mindful of at least three problematic tendencies that need to be guarded against. First is the tendency to see cosmopolitanism as a phenomenon exclusively of the present day. Thus there are problems with simply seeing cosmopolitanism as a condition of the present, for this assumes that nothing changed between Kant's early formulations and the developments of the present day. As Robert Fine (2003: 458) puts it, 'in this story, it would appear nothing substantial happened for around 300 years and then, all of a sudden, in our own times, everything happened at once'. Second, and perhaps somewhat ironically given cosmopolitanism's vaunted concern with difference, is the focus on the Western city, particularly the British and North American city within debates on the subject. The third problematic tendency concerns the neglect of questions of embodiment, gender and sexuality in accounts of the cosmopolite. Thus how these issues intersect with the sorts of formations, processes and performances detailed in the chapters above needs further investigation and debate.

Cosmopolitanism is not a panacea for the future of cities. The chapters in this volume have revealed its problematic status, multiple uses and forms of deployment, as well as the socio-spatial injustices produced despite cosmopolitanism being supposedly antithetical to intolerance and exclusion. Yet despite the fact that we and the contributors to this collection are often critical of how the term has been operationalised in urban space, we believe that it still has value as a concept and practice for the development of contemporary society. Its core values, though far from straightforward, are still relevant to how we negotiate everyday life in the city. Given contemporary developments in globalisation, migration, the exchange of ideas, multiculturalism, clashes around difference and the efforts of a particular Western hegemony in spreading a version of liberal democracy, the values of cosmopolitanism as a vision of how to negotiate and deal with difference in society have never been more pertinent than today.

References

Amin, A. (2002) 'Ethnicity and the multicultural city: living with diversity', *Environment and Planning A*, 34: 959–80.

Barnett, C. (2005) 'Ways of relating: hospitality and the acknowledgement of otherness', *Progress in Human Geography*, 29: 5–21.

Brennan, T. (1997) *At Home in the World: Cosmopolitanism Now*, Cambridge, MA: Harvard University Press.

Chan, W. F. (2005) 'A gift of a pagoda, the presence of a prominent citizen, and the possibilities of hospitality', *Environment and Planning D: Society and Space*, 23: 11–28.

Chandler, D. (2003) 'The cosmopolitan paradox: response to Robbins', *Radical Philosophy*, 118 (March/April): 25–30.

Derrida, J. (2000) 'Hospitality', *Angelaki*, 5: 3–18.

Derrida, J. (2001) *On Cosmopolitanism and Forgiveness*, London: Routledge.

Fine, R. (2003) 'Taking the "ism" out of cosmopolitanism: an essay in reconstruction', *European Journal of Social Theory*, 6: 451–70.

Robbins, B. (2002) 'What's left of cosmopolitanism?', *Radical Philosophy*, 116 (November/December): 30–7.

Robbins, B. (2003) 'Reply to Chandler', *Radical Philosophy*, 118 (March/April): 31–2.

Žižek, S. (1997) 'Multiculturalism, or, the cultural logic of multinational capitalism', *New Left Review*, 225: 28–52.

Index

agonistic politics 47–50
Aksartova, S. 12
alliances 82, 83–4
Amin, A. 124, 126, 143; *Ethnicity and the multicultural city* 42–5
Amsterdam *see* Bijlmermeer, Amsterdam
Anderton, James 230
Appadurai, A. 148
Appiah, K. 11
arts 147, 149
assimilation of immigrants 124–5, 177, 181
attitude of cosmopolitanism 7–8, 11–13, 16, 227
Australia 74; *see also* Sydney

Banglatown 26, 137–8
Barnes, T.J. 152
Baudelaire, C. 100
Bauman, Z. 71
Beck, U. 6, 13, 22, 93, 96–7, 188, 222
Benedictus, L. 1
Bennetto, J. 1–2
Berubé, A. 226
Bianchini, F. 205, 207–8
Bijlmermeer, Amsterdam 29, 184–5; ethnic diversity in 177–8, 181, 183; historical development of 179–80; urban renewal of 29, 180–1, 182
Binnie, J. 113, 133, 135, 249
Birmingham 45, 206–7; Chinese Quarter 207, 214, 216; cosmopolitan urban environment 204–5, 207; ethnic entrepreneurs 213–14, 215, 217, 247; 'Highbury 3' conference 204–5; managing cultural diversity 204–5, 207; rebranding 30, 204–5, 207
Bishop, B. 132
Bloomfield, J. 205, 207–8
Bodaar, A. 29, 247
Boho Britain Creativity Index 2
boundaries 73, 80, 81
Bourdieu, P. 226
branding 18, 19–20, 30–1, 137, 229; Birmingham 20, 204–5, 207; Spitalfields 26–7
Brennan, T. 222
Brenner, N. 232
Brick Lane 134, 137–8, 140–1
Bridge, G. 23, 248
Bristol, gentrification 62–4, 65; housing and neighbourhood aesthetics 62–3
Brown, G. 26, 249
business, cosmopolitanism in 10, 182
Butler, T. 16, 61–2, 64, 65

Caglar, A. 125
Calhoun, C. 10, 38
Canada 45, 113–14; migration to 20, 123; multiculturalism 123; *see also* Montréal
capitalism 12, 173–4, 240
Cartier, C. 148, 150

Chan, W.F. 30. 247
Chang, T.C. 149
Cheah, P. 5, 222
citizenship 30, 247, 251; denial of 194–5; global 5–7, 13, 173
city 174; criteria for cosmopolitanism 115; division of labour in 95; shared representations of 120–2
Cityside Regeneration 137–8
civic culture 40, 44, 77, 79
civic welfarism 194, 195, 200, 201–2, 248
civitas augescens 210
Clarke, D. 225
class 8–10, 11, 29–30, 198; cosmopolitanism and 187, 227–8, 235, 247, 250; cultural diversity and 193, 196–7, 198–9, 226; cultural exchange 196–7, 198–9; invisibility of 191–2; and 'Other' 172, 173–5, 183,184; sexuality and 225–6, 228; *see also* middle class; working class
Clifford, J. 166
Cohen, P. 226
Cohen, R. 112, 114
commodification 12, 13, 24–5, 173
common interest 78–9, 83
community 41–2, 55; shared commitment to 48–9, 50
competence, cosmopolitan 94, 113, 148, 223
conflicts, intercultural 26, 40, 46, 47, 122, 125
Conley, V. 54, 82
consumption 224, 225, 236, 250; cosmopolitan 24–6, 134, 173, 178, 181; gay 230
contact zones 156–9
Cook, Robin 212
cosmopolitanism 93–6; definitions 4–5, 7, 94–5, 112, 113, 173; as shared representation of the city 120–2
cosmopolitanization 22, 96–8, 107–8
creativity, diversity and 2
criminality 2
critical discourse 54
cultural capital 57–8, 131, 134, 250; circuits of 53; economic capital and 60, 65; and gentrification 14–15, 23, 58–65; reproduction of 61, 63, 67; selective use of 23
cultural diversity 4, 7; as an asset 49–50, 210, 211–15, 247; class and 193, 196–7, 198–9, 226; managing 204–5, 207, 208–10, 216; *see also* ethnic diversity
cultural exchange, class and 196–7, 198–9
cultural quarters 18–20
culture 197, 222, 226; economy and 191, 192–3

Delany, S. 143
Delanty, G. 222
Deleuze, G. 54
demands, on urban inhabitants 79–80, 81
Demos 2–3
Derrida, J. 74, 75, 206
deterritorialisation 148
Deutsche, R. 71, 78, 80
Dewey, John 66
difference 16, 40, 97, 224, 233; attitude to 102, 54, 223; of class 189; encounters with 1–2, 8, 15, 25, 251–2; exclusion of 16, 25, 28; indifference to 77; living with 66–7; regulation of 171, 175; right to 48
Diken, B. 74
distinction 14, 15–16, 241
diversity 2, 3–4; *see also* cultural diversity; ethnic diversity
Donald, J. 38, 71, 78; *Imagining the modern city* 41–2
Duncan, J.S. 152

Eagleton, T. 152
economy 148, 149; culture and 191, 192–3
education 8, 238; cultural capital and 61, 64
elites 56, 222–3; transnational 9, 10–11, 17
enclaves, elite 17, 56, 249, 250; homogeneity of 17, 20, 25; isolation of 23, 40
estrangement 76, 77–9, 80
ethics of cosmopolitanism 77–8, 82
ethnic diversity 1, 94, 95, 112, 212; Amsterdam 177–8, 181, 183; Birmingham 207, 215, 216, 247; marketing 215; Montréal 115–16, 118, 120, 124; Singapore 150; *see also* cultural diversity

ethnic entrepreneurs 213–14, 215, 217, 247
ethnicity 239; and otherness 172, 173–5, 183, 184
everyday living 78, 139–43, 251–2
exclusion 81, 82; from cosmopolitan space 113, 239–40, 248–9; of difference 16, 25, 28

Fairburn, M. 103
Family Pathfinders (Houston) 29, 196–9
Featherstone, M. 174
Fine, R. 5, 6, 252
Florida, R. 132, 221, 225
Flusty, S. 28
'from below' cosmopolitanism 9–10

gated communities 24, 40, 71
gay culture 221
gay identity 132–3, 135, 136, 142; global gay 225
gay spaces 26–7, 132–3, 134, 237–8; as cosmopolitan 135, 140–3, 144
gender, cosmopolitanism and 8
gentrification 14–17, 28, 130; cultural capital and 58–65; metropolitan habitus of 65
geography 248; cosmopolitanism and 147–8, 150
Germain, A. 26, 247, 249
global city 56
global identity 15–16
globalization 5, 43, 94, 112, 188; homogeneity and 97; in Singapore 148, 149
Goh Chok Tong 146, 150; National Day Rally speech 148, 150–1
good taste 56, 60
Gouldner, A. 54, 59
Greenwich Village (New York) 39–40
Guattari, F. 54

Habermas, J. 55
habitus 57–8; gentrification and 58–9
Hage, G. 74
Hall, P. 213
Hall, S. 46
Hannerz, U. 7, 17, 54, 57, 94, 96, 143, 223, 227
Harvey, D. 38
Haylett, C. 29, 227, 228, 247
heartlanders 147, 150–1; friction with cosmopolitans 156–9, 165; places of 162–5; rootedness of 153–6
heterogeneity 97–8, 105
Hiebert, D. 4, 5, 20–1, 113–14
Hillier, J. 58
Hollinger, D. 38
homogeneity 16, 18, 29; of enclaves 17, 20, 29
homosexuality 132, 221, 224
Hong Kong: Filipino workers in 114; migrants from 214–15
hospitality 20, 24, 30, 73–5, 210–11, 217; Derrida and 206; transforming spaces of 104–6
housing 162–4; ethnic mixture through 43–4, 177, 179–80, 182; gentrification and 62–3
Houston, Texas 29–30, 190–1, 192–3; cosmopolitanism in 199–200; urban culture in 191; workfare programmes 195–9, 201; working-class poor 192–3, 199
hybridity 9, 26, 76, 96

identity 8–9, 13, 198, 199; cosmopolitan districts and 20–1; culture and 47; ethnicity and 133–4; gay 132–3, 135, 136, 142; global 15–16; group 49
immigrants 43, 179; assimilation of 124–5, 177, 181; cosmopolitanism of 20, 174; integration of 125, 175–7, 216
indifference 77
inequality 191–2, 200, 201, 213–14, 215
integration of immigrants 125, 175–7, 216
interactions 82–3, 113, 114, 156; *see also* intercultural contacts
interconnectedness 97
intercultural contacts 40–1, 45–7, 79, 114, 208–9; ethnic cuisine and 121, 131, 214; fictional 156–9; in public spaces 118–20, 121, 156; trade and 216
interculturalism 30, 46–50, 123–4, 208–10
internationalism 150
Iveson, K. 23–4

Jackson, P. 225
Jacobs, J.M. 19
Juteau, D. 123

Kahn, B.M. 73, 115, 117, 122
Kant, I. 38, 73
Keohane, K. 28–9
knowledge of cosmopolites 6, 8, 94, 238; cultural capital and 54, 67, 225, 227; geographical 147–8; professional 57, 58
Kong, L. 152
Kristeva, J. 77
Kwok, K.W. 149

Lamont, M. 12
landscape 149, 160–5, 166
Latham, A. 25, 248
Latouche, D. 121
Laurier, E. 82, 83
Law, L. 114
law, cosmopolitanism and 5–6
Lefebvre, H. 137
Ley, D. 7, 9, 14, 16
Lim, C. 153; *Following the wrong god home* 157–8, 162, 164–5
literature 147, 151–2; contributing to environmental knowing 152; Singaporean 151, 152–3, 152–6
Litvak, J. 223–4, 225, 236
local 96–7, 248; pathologisation of 15–16
local state 26, 29, 78; development of cosmopolitan enclaves 18–19, 28; in Houston 191–2; and immigration 215
localization 62, 64, 67
locals 54, 94
loft living 58
London 1–2, 16, 19; gentrification on 61–2, 64–5; *see also* Spitalfields
looking, ways of 101

Manchester 3; cultural entrepreneurship 230–2; gay tourism 232; Gay Village 30–1, 113, 133, 220, 228–32; cosmopolitanism of 233–4, 239, 240; straight women in 236–9)
marketing 172, 173, 184, 224, 232, 250; of diversity 131, 177, 215; of gayness 234
masculinity, changing conceptualizations 136
Massey, D. 223
May, J. 16
media consumption 12
Mellor, R. 230
metrosexuals 136, 140
micropublics 26, 44–5, 78, 124, 248
middle class 11, 132, 197; new 14–17, 65, 250
Mills, C. 58
Mitchell, A. 103
Mitchell, K. 131, 177
mobility 10–11, 16, 54, 76
Montréal 26, 112–13, 123; as cosmopolitan city 115–16, 120–1, 122–6; immigrant population 116–17, 124; Mile End 120–2; multiethnic neighbourhoods 118–20, 121–2; sociability in public spaces 118–20
Mort, F. 18, 223, 224
Mouffe, C. 46
multiculturalism 123, 131, 171–2, 184; and hospitality 74; in Netherlands 175–6, 181, 183, 185; politics of 46
multiracialism, in Singapore 150
mundane cosmopolitanism 12–13, 135, 140–3, 249

nation-state 6–7
nationalism 177; in Québec 123, 124, 125; in Singapore 146, 150, 151
Nava, M. 8, 83
neighbourhoods: aesthetics of 62, 63, 64–5; heterogeneity of 66–7; multiethnic 19–20, 40, 118–20, 121–2, 126, 191
neo-conservatism 7
neo-liberalism 28, 131, 192, 200, 201
Netherlands: multicultural policy 175–6, 181; urban renewal 175, 178, 180–1, 182; *see also* Bijlmermeer, Amsterdam
New Urbanism 41
New Zealand: 1960s society 102–4; cosmopolitanism in 92–3, 98–101, 104–8; reforming zeal 103; urban culture 90, 103–4, 107–8

ocular economy 101
Oliver, B. 102
openness to otherness 7, 83, 114, 249; class and 193–4; cosmopolitan attitude of 54–5, 81, 134; cultural cosmopolitanism and 65–6, 150, 173; and hybridity 77–8
Other 15–16, 178, 200; class/ethnicity and 30, 172; 173–5, 183, 184; contact with 193–4; fear of 23–4;

solidarity with 6; welcoming 210, 215, 217; *see also* openness to otherness; strangers

Parekh, B. 50
Parker, D. 211
Pécoud, A. 10
Personal Responsibility and Work Opportunity Reconciliation Act 1996 (USA) 195
Planning for the cosmopolitan city 205, 207–11
Pocock, D. 152
Podmore, J. 58
Pollock, S. 217
power relations 8–9
Pratt, M.L. 156
professional class 9, 54
property development 14
public sex sites 142–3
public spaces 115; cultural interaction in 118–20, 121, 156; in Montréal 117–18, 124, 126; sociability in 118–20

Québec 123–4; French language in 124; interculturalism in 123–4; nationalism in 123, 124, 125

Raban, J. 71
race riots 42, 43, 44
racialization 199
racism 20, 43, 44
Radice, M. 26, 247, 249
rationality 55, 66, 67; transversal 53, 54–6, 64, 67
reasonableness 78, 81
regeneration 29; ethnic diversity as driver for 131; of Spitalfields 130, 137–8
Relph, E. 153
representational space 138–9
responsibilities 6
rights 45, 48
Robbins, B. 5, 33, 82, 174, 222, 224
Robson, G. 61–2, 64, 65
Rofe, M.W. 15–16
Rogers, Richard 41, 44
Rooksby, E. 58
rootlessness/rootedness 153–6, 158–9, 165
Rose, N. 18, 224, 232
Rushbrook, D. 134
Rushdie, S. 37

Sa'at, A., *Corridor* 162, 163
Sandercock, L. 18, 21–2, 22–3, 73, 74, 78, 79, 118, 248; *Towards cosmopolis* 70, 207–8
Schein, L. 12, 224, 239
schooling 61–2, 63; catchment areas 64; parental choice 64; and socialization 124–5
Schrag, C.O. 55
segregation 209–10
Sennett, R. 42, 71, 118; *Flesh and stone* 39–41
sense of belonging 48–9
separatism 40
settlement houses 66–7
sexual difference 26–7, 30–1, 221, 224–5; *see also* gay spaces
sexuality 133, 221–2; class and 225–6, 228; and exclusion 249; and trendiness 236
shared problems 79, 83
Shaw, S. 19
Shiau, D., *Heartland* 153–4, 157, 159, 161, 163, 164
Simmel, G. 72, 95, 118
Simpson, M. 136
Sinfield, A. 135–6
Singapore 27; cosmopolitanism in 148–51, 160–5; Housing Development Board estates 162–4; 'Renaissance City' 149; 'Singapore 21 Vision' 146, 151
Skeggs, B. 113, 133, 135, 235, 249
skills, cosmopolitan 8–9, 10, 223
Smith, M.P. 223
sociability 118–20
social solidarity 6, 39, 46, 200–1
sophistication 223, 235
space 17–18, 56; European 233; sexualized 229–32; *see also* gay spaces; public spaces
spatial practices 139
Spitalfields 132, 133–4, 143–4; gay space in 132, 134, 135, 136, 140–3, 144; immigrants in 130, 133; rebranding of 26–7, 137; regeneration of 130, 137–8
state: discourse on ethnic minorities 27, 211–13; regulation of ethnic diversity 175–7, 211–14, 250
Stevens, Sir John 2
strangers 71, 72–3, 80; community of 80; dialogue with 78, 79, 81; and hospitable cosmopolis 73–5;

hostility towards 82; opportunities for 115, 117; scapegoating 77; welcome for 73, 74, 75
subjectivity 4, 62, 193, 223, 227, 233
surveillance technologies 71
Sydney, gentrification in 15–16, 60–1
symbolic capital 56
Szerszynski, B. 4, 6, 12

Tajbakhsh, K. 70, 71, 77–8, 79
Tan, H.H.: *Foreign bodies* 155, 157, 160, 162, 164; *Mammon, Inc.* 154–5. 157, 159
Taylor, I. 230
Tham, C., *Skimming* 157, 164
Thatcher, Margaret 211
togetherness 79, 83
tolerance 115, 122, 249
tourism 6, 19, 149, 227; gay 234
transculturalism 56
transnational elites 9; in enclaves 17; mobility of 10–11
transnational networks 174
transnationalism 4, 114, 201
transversal rationality 53, 54–6, 64, 67
Tuan, Y. 152, 173, 174, 183
Turner, B. 11, 227

United Different Voices (Amsterdam) 183
urban culture, cosmopolitanism as 113–15, 187–8, 189–90, 200
urban governance *see* local state
urban planning 78, 179–80, 180–1, 182, 212; gated communities 24, 40; integration of diverse communities 29, 44; rebranding as cosmopolitan 137–8; spaces for intercultural dialogue 23, 40, 248
urbane-ness 95
urbanization 95–6, 149
Urry, J. 4, 6, 12
USA, work and welfare in 194–9, 201

Van der Veer, P. 38
Vancouver 45, 113–14; immigrant communities 20
variety 95, 115; awareness of 94
Vertovec, S. 112, 114
virtues, cosmopolitan 24, 78, 80–1, 224

watching people 100–1
welcoming others 73, 74, 75, 210, 215, 217
welfarism 194, 200–1; *see also* civic welfarism; work and welfare
Welsch, W. 55, 56
Werbner, P. 8, 9, 223
whiteness, aesthetics of gentrification 60
Wirth, L. 96
women 8
work and welfare 190; in USA 194–9, 201
working class 11, 188, 223, 235; contact with middle class culture 196–7, 198; cosmopolitanism of 132, 134, 174, 184; cultural exploitation of 199–200; geographies of 192–3; as non-cosmopolitans 16, 227–8, 236, 237, 238; as Other 30; transnational 9

xenophobia 71

Yeoh, B. 6–7
Young, I.M. 71, 78

Žižek, S. 222. 240, 250
Zukin, S. 58

eBooks

eBooks – at www.eBookstore.tandf.co.uk

A library at your fingertips!

eBooks are electronic versions of printed books. You can store them on your PC/laptop or browse them online.

They have advantages for anyone needing rapid access to a wide variety of published, copyright information.

eBooks can help your research by enabling you to bookmark chapters, annotate text and use instant searches to find specific words or phrases. Several eBook files would fit on even a small laptop or PDA.

NEW: Save money by eSubscribing: cheap, online access to any eBook for as long as you need it.

Annual subscription packages

We now offer special low-cost bulk subscriptions to packages of eBooks in certain subject areas. These are available to libraries or to individuals.

For more information please contact webmaster.ebooks@tandf.co.uk

We're continually developing the eBook concept, so keep up to date by visiting the website.

www.eBookstore.tandf.co.uk